# CONTENTS

## STATEMENTS

THURSDAY, JULY 14, 2005



## ADDITIONAL MATERIAL

# ROUNDTABLE DISCUSSION: WHEN TERROR STRIKES—PREPARING AN EFFECTIVE AND IMMEDIATE PUBLIC HEALTH RESPONSE

---

**THURSDAY, JULY 14, 2005**

U.S. SENATE,
COMMITTEE ON HEALTH, EDUCATION, LABOR, AND PENSIONS,
*Washington, DC.*

The committee met, pursuant to notice, at 10:03 a.m., in room SD–430, Dirksen Senate Office Building, Hon. Richard Burr presiding.

Present: Senators Enzi, Burr, and Hatch.

Staff present: Dr. Bob Kadlec, David Schmickel, Kira Bacal, David Dorsey, and Jennifer Romans.

OPENING STATEMENT OF SENATOR BURR

Senator BURR. Good morning. Let me call the Roundtable to order, if I can, and let me take this opportunity to welcome all of our guests.

As is the fashion on the Hill, our plans are going to change, and I want to apologize to everybody in the room. We go into a series of votes momentarily, a series that will probably last well past noon. I have asked Bob Kadlec to chair this Roundtable in our absence and I have asked the appropriate committee staff people to be available to ask questions. So we are trying to adapt to make your time as efficient as we can and to also get the valuable information that we need as we proceed further down the road to the constructive legislation.

But let me thank all of you for coming this morning to the first formal Roundtable. For the purposes of a time line, let me suggest to you that we will hold two additional hearings this month. It is our intent to be at a point where we can begin to talk about legislative language at that point. Next week, I think we will have the authors of the current legislation that has been introduced, who will testify in front of the subcommittee. And the last hearing, probably the last week of July, will deal specifically with surveillance. So we have covered a lot of ground in a very short period of time, and we are certainly anxious to listen to the information that you are here to provide us with today.

Clearly, I am grateful to this distinguished group of panelists who have come today to help us try to overcome the challenges and take advantage of the opportunities that confront us. My subcommittee, with the support of Chairman Enzi and Senators Frist, Hatch, and Kennedy, is committed to draft legislation that will significantly improve our Nation's ongoing biodefense efforts; now, as I see it, to be prepared for the inevitable—a disease outbreak that is deliberate or accidental or, in fact, natural.

Before inviting my two colleagues on each side to make any opening remarks they would like to, let me introduce our participants, if I can.

Ms. Bronwen Kaye, Senior Director of Government with Wyeth, welcome.

Dr. Elin Gursky, Principal Deputy for Biodefense, ANSER.

Mr. Clay Elward, Benefit Plan Design Manager at Caterpillar.

Mr. David Wright, President, CEO of PharmAthene Pharmaceutical.

Dr. Leah Devlin, whom I am proud to have here, State Health Director from my home State of North Carolina.

Mr. John Clerici from McKenna Long and Aldridge.

Mr. Chuck Ludlam, former Legal Counsel to Senator Lieberman. Welcome back, Chuck. You never thought retirement would be like this, did you? Just a different side of the table, right?

Dr. David Franz, Director of the National Agricultural Biosecurity Center at Kansas State University.

Dr. Tara O'Toole—welcome back, Tara—CEO, Director of the Center for Biosecurity at the University of Pittsburgh Medical Center.

Dr. John Pournoor from 3M Corporation.

Mr. George Barrett, President and CEO of Teva Pharmaceutical.

And finally but not least, George Conk, Adjunct Professor for Fordham Law School.

At this time, I would like to recognize Senator Enzi.

The CHAIRMAN. I defer to Senator Hatch, who was here before I was.

Senator HATCH. Oh, no, that is fine. I defer to you. We are always happy to defer around here to our friends.

## OPENING STATEMENT OF SENATOR HATCH

Senator HATCH. Let me thank Senator Burr for convening this and, of course, holding this meeting as part of a series of hearings on the area of public health preparedness. As you all know, this topic is one of great concern to the three of us here and, of course, Senator Lieberman as well.

Senator Lieberman and I, joined by Senator Brownback, introduced S. 975, the project BioShield II. And as we will hear, this bill provides a comprehensive approach to engaging private enterprise in the area of bioterror prevention and countermeasures.

I appreciate the committee's desire to ensure that we will pass the best possible bill by investigating all of the possible options. I want to say on a note of caution, however, that this issue is just plain too large and too serious for us to just nibble around the edges. Only broad, sweeping, innovative approaches will allow us to realign the Public Health Service, develop the necessary medical countermeasures, and protect our agriculture.

Now, this is my belief and I believe it is shared not only by the cosponsors of BioShield II, but also by the groups that have endorsed our bill, including the Infectious Diseases Society of America, the International AIDS Vaccine Initiative, and of course the

American Society of Tropical Medicine and Hygiene. To me, this is an extremely important bill and it is extremely important that we have the best input and ideas that we can possibly get.

I want to particularly thank the members of the staff of the various Senators involved—Mr. Chuck Ludlam here, who is more than a good advisor. I have to say he has been around here a long time and understands this place very well and has been very helpful. And I want to thank all of you for being willing to participate and assist us here in this understanding.

We are sorry that we have these votes so the members are going to have to be over there on the floor. But we are going to pay very strict attention to what you folks suggest to us and hopefully we can all come together and do what is right for our country.

Thanks, Senator Burr.

OPENING STATEMENT OF SENATOR ENZI

The CHAIRMAN. Well, I do want to thank Chairman Burr for putting together this Roundtable and I want to thank this great group of participants from very diverse backgrounds that can supply us with answers to at least three key questions that we posed. We found the Roundtable format to be a quicker way of gaining more knowledge and involving more people. And one of the ways of involvement, of course, would be through written questions that we will have in some other areas of your expertise that we hope that you will respond to following the hearing. This helps us to build a body of knowledge that is very useful when we come to drafting the legislation, and probably even more critical when it comes to selling the legislation. None of it is worth anything unless we get it finished, although there are a lot of spinoff ideas that come out of roundtables such as this, that often can be put into effect even without legislation. So you will have some short-term effects, you will have some long-term effects.

We do appreciate all of you participating in it. The attacks last week in London served as a reminder that there is some terrorism out there yet and there is an important job to be done. And while we pray for the victims of the atrocity and their families, we also have to strengthen our resolve and ensure that we do the right thing in all of the areas of terrorism that will keep this country safe.

We have made some remarkable strides, but we have to identify and address our Nation's weaknesses in regard to the biological threats. We know that there is a lot that still has to be done. At present, our pharmaceutical industry is not commercializing enough drugs to fight infectious diseases, whether they spread naturally or through the intentional or accidental efforts of man. Last week the New York Times reported an outbreak of influenza in Cambodia that inundated hospitals with thousands of infected children. In this country, the rise in incidence of antibiotic resistant infections is troubling and demands our immediate attention.

As I have said before, the lack of productivity in this area is not a market failure but a reaction to the incentives that encourage companies to allocate resources to tackle chronic diseases instead of infectious diseases. We have to enact legislation now that encourages new resources to be allocated to address the threat posed for infectious diseases.

Mr. Chairman, I would ask that my full statement be included in the record. It will save a little time.

Senator BURR. Without objection.

[The prepared statement of Senator Enzi follows:]

PREPARED STATEMENT OF SENATOR MICHAEL B. ENZI

Thank you Mr. Chairman for holding today's Roundtable on the scope of our biodefense bill. I look forward to continuing to work with you as we lead the HELP Committee in the work that must be done to craft a bill that will enable us to respond effectively and immediately to a biological outbreak or a bioterror attack.

The attacks last week in London served as a harsh reminder of the importance of the job that is before us. While we pray for the victims of this atrocity and their families we must also strengthen our resolve to ensure we are as prepared as we can be for an attack in this country. Although in the past couple years we have made remarkable strides in the effort to identify and address our Nation's weaknesses with regard to biological threats, the fact remains that more must be done. More countermeasures are needed to protect us from infectious disease. The public health system needs to be strengthened and we need to ensure our Nation's food supply is safe and protected from harm. Our bill must address all of these challenges.

At present, our pharmaceutical industry is not commercializing enough drugs to fight infectious diseases—whether they are spread naturally, or through the intentional or accidental efforts of man. Last week the New York Times reported an outbreak of influenza in Cambodia that inundated hospitals with thousands of infected children. In this country, the rise in the incidence of antibiotic resistant infections is troubling and demands our immediate attention. As I have said before, the lack of productivity in this area is not a market failure, but a reaction to the incentives that encourage companies to allocate resources to tackle chronic diseases instead of infectious diseases. We must enact legislation now that encourages new resources to be allocated to address the threat posed by infectious diseases.

When the public health system was conceived and developed, bioterrorism was not even a remote consideration. With an appropriate infrastructure and information technology systems, infectious disease can now be tracked and addressed. The public health infrastructure can grow to help us with bioterrorism and infectious disease.

Each weekend when I travel to Wyoming I see miles and miles of beautiful farm land that is vulnerable to attack. That vulnerability needs to be addressed, too.

When it is written, the committee bill will include provisions to encourage product development, strengthen the public health care system and protect our food supply. This Roundtable and the hearing next week will help shape the scope and terms of our bill.

In this country, we are blessed with the resources we will need to respond to these challenges. Our economy is very dynamic. We are home to some of the greatest minds in the world, and industry in this country has risen to meet similar challenges before. This committee has harvested our American power to innovate before. For instance, the pharmaceutical and biotechnology industries responded to the orphan drug legislation championed by my colleagues and friends Senator Kennedy and Senator Hatch. The challenge of that legislation resulted in treatments for diseases like multiple sclerosis, where no therapies had existed before. It may be that more of those incentives are needed, or that a new paradigm is needed with a different role for government to play in the battle that lies before us.

Again, I thank Chairman Burr and each of you for coming here today to engage in this Roundtable discussion of how we might best address the challenges that lie before us.

The CHAIRMAN. I defer back to you so you can get started on gathering information.

Senator BURR. Thank you, Mr. Chairman.

Before we begin, let me just sort of give you the ground rules. At about 10:30 I will adjourn from the formal nature of this Roundtable for what I understand to be Senate procedural reasons, at which time the staff will take over and we will take your answers in the same way we did if we were in a formal Roundtable.

We have provided for each of you three questions, and I think, for the most part, we have received answers to those. I will try to delve into some specific areas that I think we need further guidance and knowledge on. I think staff will continue on that format. These questions will be thrown open to our entire panel. If you feel the urge to respond, which I hope you will, if you will raise your hand. I was going to have you turn your cards, but I see a piece of tape on your cards, so that might be a little more difficult. So if you would just raise your hand, we will call on you.

And if I could get under way specifically as it related to Question 1, which was: What additional incentives or other measures will ensure the timely availability of sufficient amounts of effective biodefense medical countermeasures, and is the cost of such incentives acceptable?

Let me focus on liability, if I can, for a second. What are our options for liability protections, and should liability protections for countermeasures be something that we consider on a case-by-case basis versus a statute that covers everything?

Chuck.

Mr. LUDLAM. Mr. Chairman, I think it is an absolute minimum for BioShield II, and I mean a minimum, to cover the liability issue. And I think what you have to say to the companies is that if they enter into a contract to produce something, risk their capital to produce a product that we need either for infectious disease or bioterrorism, they will absolutely get liability protection without any doubt whatsoever. You have to tell it up front before they risk a single dollar of their capital to develop the product. And if you don't tell them that, the conversation will end.

Dr. CLERICI. Mr. Chairman, if I can add to that—and I worked very closely with both Chuck and Senator Hatch and Senator Lieberman as well as Senator Gregg on the liability provisions both in S. 3 and in S. 975. That work on those bills was informed by our experience in working with companies such as Sanofi Pasteur that bid on the initial BioShield procurement, the anthrax vaccine procurement. And quite frankly, liability was the threshold issue that prevented Sanofi, or at that point Aventis Pasteur, from proceeding.

We have seen again and again companies of any sort of size—quite frankly, anyone who is publicly traded—shy away from this market because of shareholder liability issues as well as the manufacturing capacity challenges of manufacturing for a market that is unknowable and puts the company at completely unknowable risks.

These products that will be derived from BioShield will be administered without the usual battery of tests and FDA approval that a normal drug will have. They can be administered under emergency use authority having been only tested in animals, not humans. To ask a company, particularly a publicly traded company, to participate in this market without liability protections is, frankly, irresponsible to the shareholders and is an absolute—I agree with what Mr. Ludlam said. A threshold question has to be overcome.

Senator BURR. I want to go to Tara, but can I ask for a better understanding of whether, John, you and Chuck talked about two different things or whether it is the same. Chuck talked about the liability protection is a key to triggering a capital investment on the part of research and development by anybody. Yours was sort of that future liability that was focused after the approval process. Are they connected? Or let me ask more specifically, Chuck, if there was a mechanism where that cost for research and development was not incurred by the company, does that now change how we look at the liability piece?

Mr. LUDLAM. Well, as I think you know, I think that for the Government to become a bio—to set up a biotech company of its own and basically to take on either the R&D responsibility or maybe even just the manufacturing, probably it will be the most costly, least effective way to proceed. So the Government could try that. I think if the Government sets up any kind of a GoCo, any kind of a mechanism like that, it will definitively end the interest of the pharmaceutical industry in its research. They will say basically, whew, we don't have to do it. The Government is going to do it. We won't be blamed for not doing it. The Government can go off and spend any amount of money on a defense contract or model, and whether it will succeed or not is their problem; it is no long our problem politically or scientifically or medically or legally.

So I think that is an extremely risky last-gasp desperation strategy if we have tried every other possible way to get this industry to play at their own risk and their own expense, which is obviously the preferred method in BioShield I and BioShield II.

So if they spend the money and they take the risk, then they are entitled to both dramatic incentives at the end, including liability protections, because they took the risk. Because if they don't succeed, they don't get the procurement, the IP, or the liability protections. And that is the way it should be. We should shift the risk to them instead the Government trying to take it on itself.

Senator BURR. Tara.

Dr. O'TOOLE. Following your second question, Senator, I was going to say I think there are a lot of different aspects to this liability question. I think if the bill does not address the liability associated with giving a drug that doesn't go through the usual safety testing procedures in an emergency situation and giving it to large numbers of people, if the companies aren't protected from nonnegligent harm in that kind of crisis situation, that will be read by the pharma-bio industry as the Congress not being serious about BioShield. I don't think it will trigger anything, but I think it will be a bellwether and seen as a very serious signal of intent or lack of seriousness on the part of the Congress.

There are a lot of other aspects to liability even in that narrowly defined space of what happens in an emergency, such as what is the animal — all about, what are the FDA regulations for getting something through that cannot be tested in the usual clinical trial format? All of that has to be delved into and made much more transparent than is the case.

I also think, however, that you are going to have to consider some form of at least minimal compensation for people who are harmed—again, not through negligence, but who suffer harm in the event of a public health crisis when we are asking everybody, for example, to be immunized in a certain city. If you don't do that, you are going to have the kind of situation we had with the smallpox vaccine where a lot of people were reluctant to put their livelihood or their lives on the line. So I think it is a complicated question.

Senator BURR. Something like the children's vaccine compensation program?

Dr. O'TOOLE. Yes. Exactly.

George.

Mr. CONK. I think there are a lot of false alarms. I don't think that there's any significant change in the liability system that is needed. What is needed is a more considered approach to what the actual threats are. Protection from nonnegligent harm is the norm. Companies are only liable if they are negligent. That is, if they fail to produce a product that conforms to specification if it is negligently designed or if they withhold necessary information. So that is the norm.

I don't think that—let's look at the points that have been raised about emergency measures. The essence of fault-based liability is that you are liable only if, under the circumstances, you acted unreasonably. If an emergency compels us or if technical necessity or ethical necessity prohibits us from doing anything more than animal testing and releasing medicines after they have been tested on animals but it wasn't possible or wasn't permissible to test them on humans, then there is no fault and there is no sound basis for liability.

Every law student in America learns in his first year in law school, his or her first year in law school that the rabies vaccine, which carried substantial health risks and faced the patient with the choice of developing rabies if the animal proved to be rabid or taking the medicine and saving his life, but at significant risk, is not a defective product. And there is no reason to believe that a vaccine developed with animal testing would be considered to be defective and a basis for liability.

As to who should be compensated, I think those who are compelled to assume risks and those who volunteer to assume risks should be compensated. That means children who are compelled to be vaccinated we provide for. I think volunteers who do things like take the smallpox vaccine, I think participants in clinical trials who volunteer and assume risks, I think they should be compensated. As to everyone else, I don't think we should volunteer to just wildly assume—I am sorry, unqualifiedly assume costs, and I think that we should essentially leave the current system in place, making individual product-by-product, situation-by-situation adaptations.

Senator BURR. George.

Mr. BARRETT. Yes, Mr. Chairman, I think we have discussed liability, at least around the table here, in two different ways, and maybe one is liability and one is risk. I do believe that as a pharmaceutical company who is engaged both in production of generic pharmaceutical and a company that does patent-based research pharmaceutical, the issue of product liability is relevant. If we are going to fast-track the approval process, which I think is a very workable and appropriate measure to consider as we think of this effort, then liability protection would be, I think, a natural connection to this.

The other question that I heard, which was one of liability related to the production, is really one that I think is addressable and may not be in these provisions, but this is the idea of guaranteed purchases of that material that has been stocked. So I think that that would be—not just, by the way, for new products but for existing products that are in the system, a guaranteed purchase of material that was prepared for this effort, I think, would be appropriate and would go a long way in encouraging us to work on this.

Senator BURR. Let me assure you, I feel certain you will field questions as it relates to the current procurement process and the clarity of understanding of it today.

David, I am going to allow staff to come to you. I have to formally adjourn this Roundtable. It will resume in an informal capacity when Bob Kadlec steps to the table.

Let me once again thank you on behalf of all of the members of the subcommittee and full committee. This may be the single most important thing we do, and we have a lot of things that are going to come out of this committee this year. Some of you have volunteered to be here multiple times, and I can't thank you enough for that. I think it displays just how difficult the task is for us to craft something that addresses a robust willingness to participate, the protection that one needs if there is a difference between public and private companies, and the recourse, George, that I think you expressed. And the question is, can we find a way to encompass all of that. And that is certainly the first intent.

If that can't be done, then the decision has to be made where do you begin to pare back and find the balance that we need. And to do that in the context of not knowing in the future what the threat is, but also not knowing whether we do this in a period of time that has a sense of urgency of a matter of years or a matter of months or a matter of weeks to be able to produce something is, in fact, a target that is ever-moving.

So I can't thank all of you enough for your willingness to be here.

The formal Roundtable is now adjourned.

[Whereupon, at 10:25 a.m., the formal Roundtable was adjourned.]

Dr. KADLEC. If I could just for a moment here invite my fellow staff members to come forward. If we could just limit it to one from each Senator's office. We will make brief introductions so you know the people asking the questions. And I do want to pick up—apparently there was one other—oh. We had one other comment to make on the liability, and we will continue that theme of questioning from my colleagues here.

Senator Burr was kind enough to make a brief introduction. I am Bob Kadlec, the staff director for the Subcommittee on Bioterrorism. I will turn to my left and just introduce David Schmickel, who is with Senator Enzi's staff in the Health, Education, Labor, and Pensions Committee. To his left is Kira Bacal, who is with Senator Hatch's office. David Dorsey, to my right, from Senator Kennedy's office, and Jennifer Romans from Senator Frist's office.

Do we have any others?

So if you will pick up the discussion on liability, if we may, at this point.

Bronwen, please.

Mrs. KAYE. I wanted to address briefly what Mr. Conk had talked about; liability and being related to negligence. I think maybe that is the way it works on paper, but unfortunately it doesn't play out that way in the court. I would like to point out, as manufacturer of oral polio vaccine, as an example, it has been a well-known, well-established fact that oral polio vaccine in about 1 in every 3 million recipients will cause a vaccine-acquired case of polio. It is a known, warned-of side effect. We had a judgment in the past 2 or 3 months, $8.5 million against us. We have had $10 million judgments. There was clearly no negligence involved and there was no design defect.

I think another unfortunate thing that happens with products like vaccines, which I think everyone views as a cornerstone of biopreparedness, is that they are given across a broad population and one suffers liability costs even for occurrences that are not connected to vaccine just because they are given to an entire population, and inevitably certain bad things will happen to certain people at any given point in time. And if it occurs in a temporal relationship to vaccines, vaccines get blamed.

I would like to offer a slightly different perspective here since we are talking about vaccines as a cornerstone, that we have a very tenuous ability in this country to manufacture vaccines. There are only four companies that manufacture the routinely available vaccines. Only three of those companies manufacture in the United States, and only two of them are U.S. corporations. And that precarious situation has the potential to become even more tenuous because of the liability burden that we are facing right now, which, by the way—going to this negligence issue—has to do with an allegation where scientific evidence has already shown that the vaccine does not cause this particular injury. And yet, we are facing a crushing litigation burden.

I think if we want to have companies that are available and willing to make vaccines, it is not a viable business to be an only-bioshield type situation; you have to be in the commercial market as well. It is too expensive and too difficult to make vaccines to not have a commercial side of your business. And the commercial side of the vaccine market is under threat from liability right now.

Dr. KADLEC. Mr. Wright.

Mr. WRIGHT. Thank you. Just two quick points. No. 1, I think this applies not only to vaccines, but to therapeutics. And it does because whether or not the current system works—and I won't get into that, but there has been a lot of debate whether the current system is even working—the fact is the current system is based upon known events. With the products we are talking about, the testing is much less than will occur on any product that reaches the market through commercial avenues.

There are going to be—there has been proposed that safety tests for an anthrax therapeutics will be around 300 subjects. There has been proposed that the safety test for an anthrax vaccine will be somewhere around 2,000 to 5,000 subjects. This is a fraction of the number of people who would undergo testing if these products were normal commercial products. And it is for the reason of it is unethical to put a product in a patient who hasn't got a condition. And so when you treat these subjects that aren't patients, you have to do lower numbers.

So there are going to be side effects and there are going to be conditions which we are not going to know about until these are used in mass populations. And that liability has to be protected against. And a company cannot afford, especially a major company cannot afford to bring a product to market with that kind of liability facing them.

Thank you.

Dr. KADLEC. Mr. Conk.

Mr. CONK. I think the problem there is that companies don't accurately perceive what the law is. This whole area of law had its origins in the asbestos cases, where virtually no testing was done. Here, what you have is a highly regulated environment in which publicly determined necessity determines that only limited testing can be done. In such cases there is no fault and there is no basis for liability.

Regarding the polio vaccine, it is absolutely correct that in a small number of cases people get polio. We don't use that vaccine anymore. Five years ago, we came to the belated decision that we should have made 20 years earlier, when France did, to stop using that design and to stop using live polio vaccine—and to use the enhanced injected killed virus which was adopted in Europe in 1968—and which we persisted in using for over 20 years.

And so if you have a handful of people every year who end up gravely injured for life and the legal system compensates them for that enormous loss, I submit that that is just fine.

Dr. KADLEC. Mr. Clerici.

Dr. CLERICI. I learned before my first day of law school that in America anyone can sue anybody for anything at any time, and they do.

Dr. KADLEC. And lose.

Dr. CLERICI. Well, but in the process of losing, these companies are losing—certainly Ms. Kaye's company and other similarly situated companies are spending upwards of $50 million in defending these needless and, quite frankly, baseless lawsuits. And remember, in the context of this question we are not talking about entering a market that is particularly profitable, that is particularly a broad market that people want to get into. We are talking about this question of liability in the terms of incentives: How do we incentivize companies that are responsible, that are capable of entering this market.

I agree 100 percent that if you are coming into a market where your profit and your market are both unknowably large, then perhaps that risk is better borne by the company that is taking that market choice electively. That is simply not the case with the biodefense market. And if we want the best and brightest companies in the world to participate, we need to incentivize them and protect them against this unknowable liability.

Dr. KADLEC. If I may, Chuck.

Mr. LUDLAM. Just to follow up on the same point, we are dealing with an industry that has essentially no interest in this market at all to start with. Basically none. I was in one of the large pharmaceutical firms several months ago, one of the only ones left with an infectious disease division at all. And they were hanging by a thread. As soon as a merger happens, they are gone. They are finished. That is what has happened in a number of other firms.

Now, the vaccine industry has basically been destroyed. We basically don't have one at this point. We have essentially nothing in the pipeline for antibiotics, and we are facing an antibiotics resistance crisis even without bioterrorism. As a former Peace Corps volunteer and a future Peace Corps volunteer, I am going to see people—I have seen people die of infectious disease and I will see more of them in Senegal when I get to Senegal. And we don't have products for them. We don't have an antiviral that kills the AIDS virus. And until we do, millions more will die.

That is the problem. And to recreate almost from scratch in the face of massive disincentives to play in this space, we need to take unbelievably aggressive actions which, at an absolute minimum, include liability protections and all of the rest of the stuff which I assume we will get into in the discussion. And if we don't do it, we will be facing infectious disease and bioterrorism forever on a scale that could be absolutely ruinous to the economy, ruinous to the civilizations. And that is the threat that we face.

So it is time to get serious, and liability protection is an absolute bare minimum.

Dr. KADLEC. Tara O'Toole.

**O'TOOLE, TARA, M.D., MPH, CEO AND DIRECTOR, CENTER FOR BIOSECURITY OF THE UNIVERSITY OF PITTSBURGH MEDICAL CENTER; ELIN GURSKIY, M.D., MSPH, PRINCIPAL DEPUTY FOR BIODEFENSE, NATIONAL STRATEGIES SUPPORT DIRECTORATE, ANSER; JOHN M. CLERICI, MCKENNA, LONG, AND ALDRIGE; GEORGE BARRETT, PRESIDENT AND CHIEF EXECUTIVE OFFICER, TEVA NORTH AMERICA; CHUCK LUDLAM, ESQ., FORMER LEGAL COUNSEL TO SENATOR JOSEPH LIEBERMAN; DAVID P. WRIGHT, PRESIDENT AND CHIEF EXECUTIVE OFFICER, PHRMATHENE; CLAY ELWARD, BENEFITS PLAN DESIGN MANAGER, CATERPILLAR INC.; LEAH M. DEVLIN, STATE HEALTH DIRECTOR, NORTH CAROLINA DEPARTMENT OF HEALTH AND HUMAN SERVICES; BRONWEN A. KAYE, SENIOR DIRECTOR, GOVERNMENT RELATIONS, WYETH; GEORGE W. CONK, ADJUNCT PROFESSOR, FORDHAM LAW SCHOOL; DAVID FRANZ, DIRECTOR, NATIONAL AGRICULTURAL BIOSECURITY CENTER; AND JOHN POURNOOR, 3M CORPORATION**

Dr. O'TOOLE. If I could echo that a little bit. You know, the chairman said that these issues are very complex and you can't solve the problem of biosecurity in one bill. And that is certainly true. Although I can understand a desire to kind of break the different parts of our security into modules and address them one by one—take on liability, for example—I think there is an urgent need to talk and think in strategic terms and to put before the country in these bills what is at stake, as Chuck said, and what we are really doing here.

There are public implications for how we handle liability that go way beyond legalistic claims and financial considerations. We need to find a way to signal to the public that if we use these countermeasures it could well be in a time of national crisis when the existence of the country is at stake. We are going to be using medicines and vaccines that have not been tested in ways that the American people have come to expect. And we have to signal to them way in advance that that is the case, that the Government understands that this is going to be a lot more risky than oral polio vaccine and yet we think it is a good idea to take this pill or accept this immunization. And that is part of what liability is going to have to do.

We are also going to have to invent whole new procedures in FDA for figuring out what is sufficient testing. There is a huge gray morass in terms of what is safe enough. I mean, you know, you are talking like a lawyer. Speaking like a physician, we don't know nuthin'. We are going to be guessing at what is good enough and hope that the benefits outweigh the risks. For some people that won't be the case.

But the public messages we are sending about the overall import of biosecurity and biodefense are as important as the specific parts of these bills.

Dr. KADLEC. Let me just follow up with a question, I think, which really was the opener that Senator Burr raised and asked: What are our options? I think it has been highlighted that it seems like comprehensive liability seems to be the suggested approach, if you will, for the manufacturers. I think it was mentioned, also, the notion of protecting those who provide the product as well as providing liability coverage.

Is there any reason to believe that you can create a case-by-case liability provision that would make sense? Or what would be, if you will, the ying and the yang of liability options as it is considered today?

Mr. CONK. Well, let's look at a couple of epidemics and see how they were dealt with in the legal system. One that I am personally familiar with is a small group of companies that were completely immunized by what is called the blood shield laws, and those were the companies that developed the concentrated protein that substituted for the genetic defect in hemophiliacs. They had complete immunity. All their patients got hepatitis and/or HIV, and virtually all of them died. And no liability burden was borne. There was a token settlement in product liability litigation of $100,000 a claimant and there was a compassionate payment afforded by the Congress.

I count that up as Exhibit A on why complete immunity from liability should not be afforded. The pasteurization of those products was perfected within 12 months of the time that the alarm went off when AIDS was identified.

Let's look at a more recent epidemic, one that didn't happen. That was the epidemic from complications of vaccinia on the smallpox issue. Now, I think that that has certain strengths. And the strength of that compensation was based on the fact that we asked, virtually required, health care workers to assume a risk to their own health in the face of no identifiable risk at all. It turned that there wasn't any risk. And people didn't volunteer. So I think that the compensation nonetheless that was afforded was a good idea and could help if properly handled, encourage people to volunteer for such efforts.

And I do think that the point regarding public education of the inevitability of risks is key not just for these medications but for all medications, that FDA approval is not an on-off switch, it is simply a stage in our study of a drug and its effects on the human organism.

Dr. KADLEC. Mr. Clerici.

Dr. CLERICI. I think your question, Dr. Kadlec, was what is the current regime—or what is the one-up regime that we are living in today. There is a statute on the books which has been used in this context. The supplier of the smallpox manufacturers, the companies that donated the old existing vaccine following the events of 9/11, received indemnification under a statute that is over 50 years old, Public Law 85–804. And that is, quite frankly, the way that the Government has approached liability to date, because there is no substitute for Public Law 85–804 that Congress has addressed. It doesn't work. The predictability of whether or not a company is going to receive indemnification from the Federal Government is not enough to incentivize the company to scale up a manufacturing facility, to dedicate its production and its opportunity costs, if you will, to a biodefense market. It just doesn't happen.

The second problem with that is one of pure fiscal concerns. In the political environment we are living in, Public Law 85–804 puts the Government on the hook, if you will, so the Government steps in the shoes of the manufacturer and itself faces unknowable and unlimited liability outside of the appropriations of Congress. We have seen again and again the Office of Management Budget refuse to grant this authority and, quite frankly, it simply doesn't work as an incentive to get companies to participate in this market.

Mr. DORSEY. If I can shift gears just a little bit. The definition of countermeasures includes products that affect or would treat the side effects of any countermeasure. And many of these many, many vaccines or treatments have very common side effects, like headache, you know, upset stomach, things that people encounter all the time and for which there are extremely large commercial markets now.

So it is conceivable, given the way some of these proposals have been drafted, that products with very large commercial markets would be—all of the incentives that these proposals provide would be available for all of such products. So we might see a waiver of product liability, for example, with respect to a vast majority of the largely commercialized drugs on the market. Is that an acceptable approach?

Mr. LUDLAM. BioShield, at least as we have drafted it, only applies to what the Government chooses to procure in that market. It doesn't apply outside of that context in the way that both are drafted. There is nothing that applies to the vaccine market generally. Maybe it should. It doesn't apply to the antibiotics market generally unless the Government procures it under BioShield. If the Government decides that that side effect is sufficiently serious, that it wishes to deploy the scheme in BioShield in favor of that and whatever that entails in terms of liability or tax, IP, whatever else, it can do it. But I think you have to think it is quite unlikely, if there is a substantial commercial market, that they would feel it necessary to deploy BioShield in order to secure the development of that product at the industry risk and expense. They could, but as a practical matter I think it is extremely unlikely.

Mr. DORSEY. But to follow up, certainly it makes sense that the Government would want to stockpile such measures for treatment in the case that they have to deploy a countermeasure.

Mr. LUDLAM. Maybe.

Mr. DORSEY. So under your proposal, there would be—you are suggesting that there should be product liability protections for a product that, arguendo, has gone through the complete FDA approval process, no just studies in animals, studies in humans, we have a lot of knowledge about its use in the marketplace because it is a commercialized-use product. What is the justification for liability protection then?

Mr. LUDLAM. Well, if the Government decides that it—I mean, the BioShield is about getting things done that would not otherwise get done. The only purpose of it is to involve the Government to secure the development of products that are not going to be developed for some other reason already without the necessary tools of BioShield. Now, if the Government—if this is going to be developed anyway because there are lots of dual use, I don't see any reason to deploy BioShield. And certainly wouldn't need to stockpile it. If there is just a substantial regular commercial market, there is no reason for the Government to stockpile it. It will exist in the pharmacies around the country and they can use it.

Mr. DORSEY. Well, I mean, as we learned with Cipro, there was a substantial commercial market for Cipro, but there weren't adequate supplies of Cipro available at the time to respond to the anthrax threat that we experienced in October of 2002. I mean, there were stockpiling issues with an otherwise commercially available product.

Dr. KADLEC. Tara, if I may, and then Mr. Wright.

Dr. O'TOOLE. I think there is a big difference between covering vaccinia immunoglobin, which is probably the classic case of something that would be used to treat a side effect of the smallpox vaccine, and covering aspirin or something else. I think the test of what gets covered under a liability rule—and I think there should be one rule, should be something other than did we acquire it for the stockpile. I think that should be a kind of product-by-product decision made at the time we start supporting the product as early in production as possible and there is one liability coverage for everything. But I think part of the, I think this ball got rolling because of the VIG issue. And, you know, ought to be more narrowly construed.

The problem with Cipro is if we are going to be using it for conditions for which we do not have much scientific evidence. So you may have a product that is in wide commercial use for many conditions other than inhalational anthrax that you are now going to use in a context that we have very little data on. That probably needs some kind of coverage.

Mr. DORSEY. Is the data we don't have data on efficacy or data on safety, in that context? Isn't it mostly on efficacy?

Dr. O'TOOLE. Both. I mean, you know, clearly people with inhalational anthrax are desperately ill very quickly.

Mr. LUDLAM. I have to add that there Cipro is the perfect case for some of the larger points here. That was the case where it wasn't procured for BioShield. They had some stockpiles, I presume. But it wasn't used for that. Unlabeled, because Bayer, at their own expense had gone forward and gotten a label for anthrax at the Government request. It was the stupidest thing Bayer ever did. Because as soon as we got hit with the anthrax attack, the Government said—Bayer donate 4 million courses of Bayer (sic) to the Government—donated. And the Government, we would like to buy 2 million doses. And by the way, if you don't give it to us at one-fourth your market price, we might challenge your patent. And that came partly from the Hill and partly from HHS.

Now, the Government had no basis for challenging the patent, but it would have tanked the stock price of Bayer. It was their lead product. And then every other person who bought Cipro came in and said, we want it at one-fourth the market price. Now, Bayer has never gone public with the damage that was done by this little incident. But it has plagued our ability to engage this industry in this research. They understand if they have the perfect product, in the middle of an attack, whether there was procured under BioShield or not, the Government will steal it. That is what they believe.

Now, that is why we have to overcome—that is the suspicions we have to overcome in BioShield II. And that means liability, tax, patent, procurement, the lot. Because we are dealing with uninsurable risks, political risks that a senior Government official will play a mafioso tactic against a company and tank their stock price.

Dr. KADLEC. David, if I could, I would like to get Mr. Mr. Wright.

Mr. WRIGHT. I think we are getting into many issues here. I think the first step, if we want companies to develop products for BioShield, there has to be protection, liability protection for those products approved under the animal rule. All right? And that is a simple first step.

The issue of Cipro is much more complex. And I totally agree, the Government cannot steal products from companies. And that has kept a lot of investment out of this industry. Our company was almost not financable because of that and because of what was expected that could be done if we spend all this venture capital money developing these products and then all of a sudden someone comes along and takes them away.

Products, though, that are on the market, that have gone through human testing, who are used on label, the liability issue is different from those products who are being used either off label, at the Government's request, or products who have not been through the standard FDA approval and have been approved by the animal rule. And that is where the first step needs to be taken, in my opinion.

Thank you.

Dr. KADLEC. Mr. Elward.

Mr. ELWARD. Well, I think, you know, Caterpillar has to face the risk-reward in every product we develop, and I don't think this is any different. Whether or not we have protection out there is just going to affect, you know, to Mr. Ludlam's point, we are going to have to have higher rewards in place. So there is a continuum there. And one size never fits all. That is why we will have to evaluate it on a case-by-case basis, just like any commercial industry does, or business does out there, look at it on a case-by-case basis. And if the protections aren't there, then we are going to have to have higher incentives in place at the end, you know, at the end of the day to get the action that we want.

Mr. DORSEY. Mr. Conk suggested that companies that are protected from liability don't have an incentive to exercise due care in producing or manufacturing their product. That was the example suggested by the blood products he mentioned. If we afford liability protections to companies, how do we ensure that we do in fact exercise due care when they—the best care they can, when they make a countermeasure?

Mrs. KAYE. The swine flu situation was another example, an analogous situation where the Government stepped in and provided liability protection for manufacturers because manufacturers were unwilling to produce that vaccine in any other scenario. And in that instance, the Government had the right to subrogate claims, so if there was some negligence on the part of the manufacturer, the Government could turn around and do that. I actually think that is a far better way to deal with these situations than throwing this open to regular plaintiffs and juries.

Getting, again, back to this polio situation, it is a very analogous situation. That is a vaccine that the U.S. Public Health Service said needed to be used in this country. It is the only effective vaccine when there is circulating wild polio virus. And the U.S. Government said this is the vaccine you must use in this country. We as a company manufactured that vaccine according to specifications, and suffered a lot of liability on account of it. It is a very analogous situation. No negligence involved.

I absolutely think that it is important to have a standard across-the-board liability system in place, product by product, it can be decided whether this product should have this system available to it. But I hope what you are not suggesting when you talk about should this be on an ad hoc basis is that different products would have different levels of protection, because then you would never know what you were getting yourself into when you started down the path.

Dr. KADLEC. John.

Dr. CLERICI. There are legislative proposals being floated now that have a bad actors exception to this, and I think that that would probably be acceptable to the majority of folks that I work with. In other words, if there is clear proof of gross negligence or willful misconduct, companies are more than happy to accept that sort of exception to any sort of liability regime, because, you know, the responsible company is not going to be willfully negligent or commit gross misconduct.

Dr. KADLEC. Mr. Barrett.

Mr. BARRETT. Yeah, I just want to make a comment about the Cipro discussion. You know, we have talked—it started with a discussion, I think, as to whether or not you can carve out, from a liability standpoint, the treatment, the use when it is in a countermeasure versus the broader use. And I think there are people at this table who are far better equipped to answer that legal question than I am.

I just want to make sure that we—because I heard the Cipro discussion as the example here and I am not sure that we don't take the wrong messages from that Cipro example. I don't know if we are going to come back to it—I hope we do. But there are ways of ensuring there was information available on other drugs that could be used as treatment for anthrax. We did eventually—Secretary Thompson did respond, after this occurred and information was circulated to the medical community, there is work that can be done ahead of time to look not just at how do we get that drug into the system, but are there other possible treatments. And again, I think that it is going to be important for us, as we think about Cipro, to take all the lessons from it. And I am not sure that I would necessarily draw the same conclusion that I heard earlier.

Dr. KADLEC. If I may at this time, maybe we can shift gears slightly. I won't say we answered all the issues on liability, but certainly what has been discussed here has certainly, I think, moved our thinking and understanding of the number of issues associated with it very forward.

But if I could turn now to maybe, we could shift to the issue of intellectual property, something, again, not so contentious, I am sure.

Mr. DORSEY. Another small issue.

Dr. KADLEC. And I would invite my colleagues, if there are any questions on intellectual property, we could open it up at this point. Otherwise, I do have one. And really, it is a more general one, is to just kind of get a sense of the page here, what are the varying opinions on intellectual property and what should we—we, being Congress—be mindful of as we move forward with consideration of a number of potential intellectual property-related considerations or provisions, that run from wild card patent to patent restorations and others?

So with that, I will just open it up to seek comments from the panelists.

David.

Mr. WRIGHT. I think if we want to get Large Pharma involved in this industry, we are going to have to do something dramatic to make it worth their while. BioShield legislation has put $5.6 billion over 10 years into a pot to cover all the products. And while that seems like a large number, if you do the math that is $560 million a year. In most large pharmas, that is not a big enough market for them to even look at a product in. All right? They would not spend money developing. In most large pharmas today, it is $750 million to $1 billion market opportunity before they will spend their money trying to develop a product for that.

So there is somehow that we are going to have to increase an incentive for them to get involved and spend their money. It takes just as much money to develop a biodefense drug with a market cap of $150 million possibly over 3 years, maybe, if the Government decides to buy it, if they think they are going to need it. It takes a Bigco just as much money to develop that product as it does to develop a billion-dollar product. And shareholders just will not stand for it.

So whether it is patent extension, whether it is on other products for every product that is developed in that field, or something like that, and while there are those that will say that is, you know, using Peter to pay Paul, it is somehow going to have to be done. Otherwise, big companies are not going to be able to get involved.

Dr. KADLEC. Mr. Elward.

Mr. ELWARD. I would agree that we definitely need to have the right incentives in place. Caterpillar is the world's leading manufacturer of—mining equipment, and we currently provide health care benefits that cover more than 140,000 lives here in the U.S. and we spent last year over $600 million on that. Like other companies that provide quality benefits, we are concerned about rising health care costs and prescription drug costs in particular. In the past couple of years, our prescription drug costs have increased 20 percent and there is no end in sight to that.

But to help address that trend, we have encouraged our employees and retirees to opt for generic drugs when available, and they do that quite frequently. When that is out there and they know about it, 90 percent of the time they are using generic drugs. And that is a huge savings for them. When they use generics, our employees save over 80 percent of what they would have spent on name-brand drugs. And our company benefits substantially as well. It is not uncommon for us to save 30 percent on our drugs-spend when our employees use those drugs, so it is both good for them and it is good for us as an employer and allows us to continue to provide those benefits to our employees.

We do believe that Congress can and should strengthen BioShield I and we need to have certain incentives in there—product liability protection, guaranteed purchasing, tax incentives, and the like. But in the current form, we could get some unintended results. And these higher prescription drug costs for our consumers and reduced pharmaceutical access for uninsured are those risks, especially our employees, our retirees who are on fixed incomes.

So those two areas of concern: One, the legislation could actually undermine the goal of developing novel countermeasures by merely encouraging minor changes to already approved products. And two, the legislation offers patent extensions, which would delay the introduction of new generic products to the marketplace, which is so important to our being able to provide those benefits and for our employees and retirees to be able to support their families that way.

So we believe that Congress can and should implement some additional incentives to help this legislation, but we need to make sure it is equitable and that it doesn't end up negatively impacting the people that rely on this so much.

Dr. KADLEC. Mr. Barrett, and then I will get Chuck and then Tara.

Mr. BARRETT. Yeah, again, I would like to remind everyone that we are a large producer, the second-largest pharmaceutical company in the United States, measured in prescriptions produced. And we are both a generic company and we are a, again, patent-writing, filing, research-based pharmaceutical company. So I think maybe we bring a slightly different perspective to this.

We are very supportive, and I think all of us here, obviously, of some of the provisions that we have seen in both BioShield I and BioShield II. We talked about some of them this morning—product liability protection, fast track, etc, etc. Our concern with some of the intellectual property provisions, that they really are indirect and we wind up essentially shifting the cost of our homeland security into the health care system.

And so, particularly as we started to look at the wild card provisions that we have seen in legislation, our concern was that it would have these unintended and really unmeasurable consequences related to the health care system. They lack transparency, as ability to know where you are spending the money and where it is going, and proportionality.

So in particular, we thought that anything that essentially slows down the access of affordable drugs probably was bad policy. There are direct ways of encouraging innovation and there are direct ways of encouraging our existing infrastructure to make products available to us. But the use of an indirect tool usually leads to a bad outcome. If you create an incentive that is indirect, we are unquestionably going to wind up with an unintended consequence. To me, the idea of putting that consequence into the health care system rather than dealing with it as part of the national defense and the homeland security seems like questionable policy.

Dr. KADLEC. Chuck.

Mr. LUDLAM. If we don't get the products we need to deal with these pathogens, we absolutely know for certain already that millions of people will die—from AIDS, malaria, TB. We know we will have an antibiotic resistance crisis and people in the United States will die of earaches. We will be in a postantibiotic era. We know if SARS gets loose that hundreds of thousands or millions will die. If avian flu gets loose, at its current lethality rates of 55 percent, millions, probably hundreds of millions, perhaps a billion people will die. We will have Rwandas and Cambodias all over the world. The 1918 flu was 1.8 percent lethal. Avian flu is currently running 55 to 70 percent lethal.

The question is what wouldn't you do to make sure that we have products that deal with these pathogens. And I think the answer is absolutely nothing that we shouldn't be willing to do to fill this gap with these medicines.

Now, I am absolutely certain, based upon hundreds of conversations with Big Pharma people, that probably the only incentive in the bill, the bills that are pending, the only incentive in the bill, assuming you do all the rest of it, the only incentive in the bill that will work, that will actually turn their heads and actually convince CFOs to play in this space are the IP provisions included in the wild card. Phil Russell in the Wall Street Journal article on Monday said God help us if we don't get some of the Big Pharmas to play. There is nobody with more credibility in this space than he is, and more frustration at trying to get these products developed. OK.

We want to get one large pharma company in every RFP that the Government puts out to get any countermeasure for any countermeasure for any pathogen, at least one. And then we want the small companies as well, because they are obviously innovative as well. It clearly is an indirect subsidy, there is no doubt about it. The generics have one or two legitimate points to make. And the Congress has to decide whether or not that argument, which I think is classically a NIMBY argument—it is a good argument, but it is a NIMBY argument—is trumped by the larger concerns.

Dr. KADLEC. Tara.

Dr. O'TOOLE. Well, again, I think we have many compelling and conflicting public issues at play here. If you believe that biosecurity is a real problem, if you believe that a bioterrorism attack, or indeed, a natural pandemic is an existential threat to the country, then the biopharma industry becomes a critical national security resource that we have got to protect and indeed promote. We have to find a way to not only ensure U.S. preeminence in these industries, but to promote innovation. And intellectual property and the ability to make money on it is one of the ways that we have succeeded historically in promoting innovation.

And we certainly need countermeasures. And I have heard from Big Pharma companies—I am not a Big Pharma company nor am I a small biotech company, but I have heard from Big Pharma, too, that what they are most interested in is these indirect incentives, which everybody agrees, I think, are a nightmare to manage, from the Government's point of view, and will indeed be an indirect cost on the health care system. And we don't want to crush the health care system in our pursuit of bioterrorism countermeasures.

So what to do. I do not think there is any way out of this pickle unless we figure out how to rapidly accelerate the time needed to make new drugs and vaccines. Time is the real cost here. And we need to be able to predict the winners and the losers a lot earlier in the process than we can now. I think the Government should enter into a long-term very ambitious commitment to radically accelerate the production of drugs and vaccines, and figure out how to get Big Pharma and the universities and the biotech companies involved in such a measure. If we don't drive down the cost of drugs and vaccines, we are cooked, both from a health care perspective and from a biodefense perspective.

Dr. KADLEC. John.

Dr. POURNOOR. Actually, I would like to maybe blend a couple of themes that were emerging in our discussions here. The central intent of any business is to develop a sustainable business model. I think risk management, protection of patents, expenditures in areas of R&D to be able to augment what is already—and R&D infrastructure in many companies are the kind of themes we need to emphasize.

I am also a believer in the fact that the long-term sustainability of it definitely has to do with not only development of new pharmaceutical, but these patients, above and beyond needing pharmaceutical and vaccine, will need hospitalization at some point in time, may need isolation. We need to register and credential health care workers—the kinds of themes that I think you would see probably emphasized from the public health perspective here. Decontamination, personal protective equipment—all of these will carry some level of risk with them and some level of liability. Clear articulation of the boundaries, I think, are fairly important.

But in the intellectual property area specifically, to incentivize companies to participate I think three key areas are quite important. First, I would say, is the initiatives around patent reform, from first to invent to first to file. To us, it is a key area that needs emphasis so that if there are activities going on in organizations in their R&D functions which, before filing, is unbeknownst to anyone in the world—no one really would know that that activity is going because it is confidential, proprietary information—that there is sustenance in that moving forward.

Prior user rights, I think, is a theme that we would like to see emphasized. With prior user rights, you give extension and a larger domain to investments that do not carry as much risk if they do not make it to filing in time.

And also research exemptions so you can see what the tangential extrapolations from existing themes that people are doing R&D on in the long run are going to be.

To us, those three pillars, from an IP perspective, I think, are important. But again, indemnification boundaries is yet another aspect of risk, as well as, I think, the amount of R&D investments that it is going to take. Sustainability of the overall business model requires a right cocktail of the various components.

Dr. KADLEC. Mr. Barrett.

Mr. BARRETT. I want to back us up a little bit. I think we do ourselves—I don't think we take ourselves in the right direction when we frame this as a generic/Big Pharma question. The question, really, is how do we mobilize our entire pharmaceutical productive capacity around the country to kick into this effort.

And again, I just want to remind you, companies like Wyeth and Pfizer, Major Pharma, are significant producers of pharmaceutical products in this country, as are companies like Teva. Three of top five largest producers in the U.S. economy of pharmaceutical are generic companies as well. So we need to mobilize the entire system. And the question is how do we do that. And I think there are aspects of IP that should be examined. The use of wild card, as I said, for me has really significant policy implications for all of us and I think, for that reason, should be looked at with some concern.

But when we look at the development of a truly novel treatment, if it is superior to anything that is available, if it is unique and there is no other therapy, then we should look carefully at whether or not there are tools that we can use in the IP area. But it should be direct. We should be able to see what we are spending, what we are spending it on, and what the outcome is that we want. And that is the reason that I express this concern about these indirect tools.

But I think, again, we will do the most work if we ask ourselves how do we mobilize the entire system. That includes all manufacturers, wholesalers, distributors, everyone, to participate at the right time, at the right speed, and at the right scale to be effective.

Dr. KADLEC. Mr. Elward.

Mr. ELWARD. What we are talking about here is insurance policies. And nobody wants to buy insurance. I don't want to but insurance, nobody here does, but we feel like we have to, and in this case, the national interest, we decided we want to do this. And it is like air defense. We don't want to go to Boeing and spend a bunch of money with Boeing to build fighter jets if we didn't think, you know, if we didn't think we needed them and we hopefully don't have to use them.

But I can't imagine going to Boeing and saying, OK, we are going to—we want you to build the latest and greatest in avionics for the next generation of fighter jets, and to incentivize you to do that, we are going to protect you from competition on the commercial side of your business. So don't worry about the next generation of AirBus competing with you on the 777 because they are not going to come—you know, they are not going to be in the game here if you can do this.

It drives completely the wrong behavior and it is not in the best interest of consumers and competition. And that is what we are all about.

Dr. KADLEC. Dr. Franz.

Dr. FRANZ. Thanks. I would just like to underscore a point that Tara made about our ability to respond very quickly. I was involved in a study on the use of vaccines in biodefense in the last year, and I essentially concluded that, beyond smallpox and anthrax, because of the nature of the agents, because of the way they present themselves, and because vaccines by definition are prophylactic, there may be very little space for vaccines in bioterrorism and emerging infectious disease. Our system just does not respond quickly enough.

To give you a concrete example, in 1996, at USAMRID we produced the first GMP lot of RPA, the recombinant anthrax vaccine which is now the subject of the BioShield I legislation. That is almost 10 years ago. We were at that time trying to get the animal rule in place. We were talking about the animal rule in place, putting it in place in the mid-1990s. It is finally in place. There is only one product, to my knowledge, that has gone through that system, and that is pyridostigmine, a prophylactic for chemical defense.

So I believe there are areas where the Government does have control in which these processes can be speeded. It is very, very difficult, even with the animal rule in place, to get answers from the FDA about how to move these products forward into advance development.

Dr. KADLEC. Mr. Ludlam and then Mr. Wright.

Mr. LUDLAM. I am absolutely delighted at George's comments about where we can go on the intellectual property provisions. And I think actually it forms the basis for a consensus, perhaps, on these issues that look to be so intractable. Because BioShield II says, page 79, lines 19 through 24, that it must be a novel product, not previously approved ever. And on page 80, lines 1 through 4, it must be superior. The word "superior" is also in the bill. It has to be novel and superior. And no wild card could possibly be granted unless it is both novel and superior. And I think George has suggested that that is a place where perhaps intellectual property incentives might be useful.

Previous to that, it says that in addition to those flat limitations, which always apply, it can only be deployed at the discretion of HHS when they find that there is nothing available, the other factor—considered. It is a truly a last resort in the most dire cases. That is the way it is drafted. I in fact was working with some of the generic pharmaceutical firms when I drafted it, under the table, behind the scenes. They actually came to me, they actually provided me language, they made sure that I was drafting something that didn't seem inadvertent. I won't reveal who it was because I told them I wouldn't out them on the process. Very carefully drafted, and I think George's statement gives us a great way to move forward.

Dr. KADLEC. Mr. Wright.

Mr. WRIGHT. Just a comment. I think the comparison to the biodefense and the defense arena is very appropriate, because what we are talking about are incentives. And what we are talking about here is giving an incentive for a company to develop a product that they would not ordinarily develop.

Now, when the aerodefense agencies are asked to develop a new F-15 or F-16 or Stealth bomber, there is a process by which they risk none or very little of their money in developing that. There is a competition that goes forward; they are paid for that. There is then a certain set of milestones that are set up, that if they meet and develop this product, then it is brought forward, they develop it, and it is purchased. And they know up front that they are going to have a market, that X number are going to be purchased, and that they will be able to make a profit and a return for their shareholders.

This does not currently exist in this business. Pfizer, Wyeth, Roche—and I can't speak for them, but I am a cochair of the Alliance for Biosecurity, that has all those people on it and I have heard them talk. And I am not speaking for the alliance here today. But what they say is we are not going to go out and develop a product for anthrax just because we have money to spend to do it. There is no reason to do this, there is no market.

So whether it be wild card, whether it be a totally different system, I think the point that I would like to make is that something has to be done to get everyone involved. Do I want Big Pharma in here? No. I am a small biotech company. I can't compete with Big Pharma. I have nothing to gain from having Big Pharma in. But the country has everything to gain because of the millions of people and the economic devastation that could occur from one anthrax attack.

Dr. KADLEC. Dr. Devlin.

Dr. DEVLIN. From a public health perspective, we certain appreciate the issues that are being discussed here today, and we recognize that there is a governmental role to providing a sustainable, predictable market for these countermeasures. And also, the liability issues have got to be resolved.

We learned a lot in the small pox vaccination issue. The public health workforce is going to be the one that delivers the medications that we're talking about today. We have got to be strong. We have got to make a very strong commitment to the public health infrastructure or this conversation is meaningless.

We manage the IND process, which is cumbersome. We manage the adverse effects process around small pox. We learned a lot.

In the last year's flu vaccine shortage, we were very grateful that the Federal Government and industry were able to get together to commandeer the amount of vaccine that was available and manage that through the public governmental infrastructure.

That is the only way that we are going to get maximum health protection when resources are short and the need is great. And the public health workforce is also going to be the ones that manages that public information and that public fear.

When you send anthrax through the mail, that puts every person in this country at risk and on alert. And managing fear is a huge part of any management of a vaccine or antidote distribution.

So I want to say we were grateful for industry and Government at the Federal level's participation last year. And I want to say that we have a strong vaccine distribution system for children's vaccines. We do not have that for adult vaccines. And that is a real problem.

So we were kind of flying by the seat of our pants last year. We were lucky that we had a light flu season, but we do need to work that system out.

And last, is I ask for your support here on the Hill and with our Senators of the public health system so we can get these drugs to the people that need them.

Thank you.

Dr. KADLEC. David, if you have a question.

Mr. DORSEY. I have a question about the level of incentives needed to get companies to play in this area. The argument is made is that is costs $800 million to a billion dollars to produce a pharmaceutical. And we need to make the market big enough to be people to play in that.

And I am wondering if that truly is the case with countermeasures. Countermeasures, everyone is arguing , are products that are not tested in humans or in very small numbers of humans. My impression has always been that the high cost of developing pharmaceutical is mostly, at least in a majority and perhaps even mostly in the clinical trials needed to get the products to market. But we do not apparently have that with countermeasures because they will be tested in animals, very small numbers of humans.

To what extend does that reduce the cost of developing these things and therefore suggest that we do not need the kind of large incentives that some have proposed to get these things developed?

Dr. KADLEC. Mrs. Kaye.

Mrs. KAYE. I think you bring up a valid point but in terms of whether the large scale clinical trials will be the largest portion of the costs for a product, I think it will vary from product to product.

On a biological product and a large part of the expense is the scale up and the development of the process, which you would have whether or not you had large clinical trials.

On a chemical entity, what you are seeing is probably correct because the scale up on the development of a chemical entity is not as significant a part of a cost of bringing a product to market.

But I think one thing that also needs to be kept in mind is that for large companies, I think there may be some analogies to the aerospace industry, but not completely. My understanding of companies like Boeing is that their opportunities in the commercial market are pretty limited, and they are struggling.

Whereas, in the bio-pharmaceutical industry, we routinely turn down research projects that are in our pipeline that we could develop because we have to limit. We do not have endless resources, and we have to pick only certain projects that we can fund and let others go by the wayside.

So we are talking about opportunity costs in Big Pharma companies saying, well, we are going to shelve this commercial product that could have a large commercial market and instead devote our scientists to this BioShield measure. And that is, I think, something that needs to be overcome in order to incentivize large companies that have a large portfolio of possibilities out there.

Dr. KADLEC. Mr. Wright.

Mr. WRIGHT. In addressing the cost, you are absolutely right. The cost of developing a product that does not have to go through full clinical trials is less. So it may be $150, $500 million to develop a bio-defense product.

Our little company so far has worked on three products. The first product we worked at, we had spent almost $35 million of our money on and then found out it did not work.

OK, so while that is costs that are sunk, and it is this type of costs that it may take $200, $300, $400 million until you find a product that works. And then you have to develop it.

I also agree that Big Pharma opportunity costs are key, because it takes just as much time to develop this product that they may make $100 million on as a product that they make a billion on. And shareholders just do not tolerate that.

Dr. KADLEC. Mr. Barrett and then Mr. Ludlam and then Mr. Conk.

Mr. BARRETT. Yes, I think that—I want to back up for one second, because as much as I love it when somebody agrees with me, I wanted to clarify something from earlier. [Laughter.]

It does not happen often enough.

My comment about the IP is very important, that my objection to wild card is across the board, and I want make that clear. And I appreciate your observation.

So my comment really was about IP provisions that would possibly provide extensions to specific products or the restoration of specific products. But wild cards, from our view, are bad policy.

On this issue, it is important to consider whether or not direct funding of clinical work is actually much more efficient that in the direct system. And that, while it seems expensive for the Federal Government to consider the real funding of the work, it is probably considerably less expensive than some system where we are going to pay the price and some way we do not see or do not have visibility on.

So I would encourage us to consider systems in which we actually pay directly for the clinical work, for the scale-up if it is a biotech facility or whatever it may be. And that may seem on the surface quite expensive. In the long run, I think it is actually very productive. And we should really be considering it.

Dr. KADLEC. Mr. Ludlam and then Mr. Conk.

Mr. LUDLAM. I knew you would oppose—I did not say you did, I just thought I would be a gentleman.

I completely agree that if there was a way to pay through the front door, that would be the way to do it.

We are heading for a budget collapse. We are heading for an implosion of Government finances. And the idea that the Government is going to put billions and billions of dollars as the procurement for products it may never use, I think is unlikely. And in fact, one of the strategies—I will be honest—in BioShield II, who has to pay the industry in ways that do not involve discretionary appropriations alone. Pay them through tax. Pay them through IP. Pay them through liability.

That is all money. Now if we could pay them entirely with procurement and maybe liability and forego the tax and patent, fine, I think that is completely out of the question in terms of what the Government will actually pay.

Five billion six hundred thousand dollars for this entire space is a joke. It is absolutely a joke. We are spending $9 billion this year on missile defense. Now if we are spending that kind of money on these products, yes, maybe we can pay them through the front door.

Anyway, I am just being honest. We did put the IP on the table because it was a form of money that we could pay that does shift costs to the health care sector. That is true. I would say that is a small price to pay for the collapse of the health care system if we get hit with a pathogen for which we have no countermeasures. Fine.

In terms of the animal models, we are assuming we have the animal models. One of the massive problems here is we do not have the animal models. And we need a whole animal model industry.

If you go Jackson Labs or Charles River, why are they going to create an animal model for a product that may never go into animals? Because there are other reasons why the industry is not playing.

You have to have the animals on line, and alive in quantity in BSL III, IV facilities before they will start the research.

You have to have manufacturing facilities for biologics on line, ready to go, available before they will start the research.

So that is why our bill applies to research too as including animal models. It applies all of this energy, including wild cards to research tools. Because if we do not have better research tools within the animal models, the industry will never start the research for that reason along, aside from the other disincentives.

Dr. KADLEC. Mr. Conk.

Mr. CONK. Well, if what Mr. Ludlam says is true, as far as what drives Government policy, I think that is a shame.

What I would like address is the point first raised by Dr. O'Toole and the whole idea about accelerated development.

It is true that what we are talking about is the down side of biodiversity. We do not know what little scrapes of RNA in birds or cat-like animals in China or somewhere in the African rain forest are going to succeed in reproducing in humans.

But if we want to look at this whole picture, let us start with the basics. The question I think we should ask first is how does each of these things serve our public health system, which is the fundamental structure through which we are going to administer public education and all mass medical defense regardless of its origin whether criminal or naturally occurring diseases.

If we are to accelerate these processes and do with kinds of testing that may be shorter and perhaps less thorough than we hoped for in the past, then what we need to do as a corresponding measure is we have to greatly increase the amount of money that we spend in studying the biological effects of the various substances, whether biological or chemical, that we are administering to the population.

We have this idea, and this affects the liability discussion too, that when something is approved by the FDA, that it is an open gate and everything is free now. I think that we should look at the whole thing in a staged way.

FDA approval of marketing we should see as the beginning of a new stage in product development. Unfortunately, today it is a stage that I think we could describe as the mass poorly controlled experiment stage where we not take small numbers—we have small numbers of people on whom we have tested, and now we have tested it millions and do very little to see how that is carried out.

So I think that the burden on companies for stewardship of their products is something that we should increase correspondingly to the advantages of accelerated testing, etc, that have been proposed.

Dr. KADLEC. Well, thank you, Mr. Conk.

What we are going to do is now if you do not mind, is we are going to transition to question two.

I think the public health people have been quietly waiting, biding their time to address this. Just so we have some idea of the timing of this, under normal circumstances, 11:30 the formal round table would conclude. But I think given that staff members are now running the show, so to speak, we will go as long as it takes— [Laughter.]

Dr. KADLEC. I think to address the issues that we have before us. So if we could open this question two. And what I would like to do is kind of blend an issue here a little bit and that is talk about the workforce, the public health work force first, but also kind of highlight, if you will, something that I think is relevant to the bio-defense industry and that is the issue of immigration requirements and the need for providing grants in education and increased training.

So if we could kind of open that up as our segue to the public health discussion. I would like to open that first and then maybe broadly discuss what we know or do not know in terms of the status of our public health infrastructure, particularly the readiness of it and what does it mean to be ready? What are the metrics around that?

So if I could just open that up first with the workforce part, and then we will transition, if you will, to what does it mean to be prepared?

Dr. Devlin.

Dr. DEVLIN. I think I mentioned earlier the notion that it is going to be the public health workforce that has to carry out the early detection of any kind of intentional or unintentional disaster and respond with quarantine, isolation, also the distribution of the vaccine and any counter measures that would be developed, going all the way through mass care to recovery to public information.

All of that requires a well-trained workforce. And we have a tremendous variety of professionals working in teams in the communities, in the States. They are nurses. They are epidemiologists. They are laboratorians. They are environmental health specialists, industrial hygienists, all of them requiring a specific amount of expertise to be able to contribute to the preparedness effort.

It is very difficult to recruit these individuals. About 20 percent of the graduates of schools of public health actually go into the practice of public health. They are in your industries. They are in hospitals. They are in research, academic centers and so forth.

So we have also an aging workforce. Our public health workforce in general is about 7 years older than the rest of the State Government, local Government workforce.

We have nationally about half of our nurses would be able to retire, and that is our largest group of the workforce within the next several years.

So we have concerns about preparing the workforce, retaining, recruiting, all of those challenging and effective local, State, Federal response to preparedness. So it is a critical issue, and I am very glad that you have had that on your sights.

There is a bill, Senate bill 506, by Hagel and Durbin, that has been developed in partnership with Federal, State, local officials. It would be similar to the rural health act of maybe 30 years ago where there would be funding created for scholarships for recruiting students into the field from the schools of public health and also loan repayment programs.

We think this is a very important tool for the future of the public's health in terms of getting the expertise into the communities in a sustainable way.

Thank you.

Dr. KADLEC. Dr. Pournoor.

Dr. POURNOOR. Under current HRSA requirements, general readiness levels for public health are set at a minimum level of 500 per million. Being able to receive 500 patients per million in population. That is 500 of 1 percent of the hospitalization rate. So this is sort of a threshold national readiness that collectively we are working toward.

If you take that 500 per million and extrapolate what the number of individuals needed and the amount of space for this surge of patients, even at 500th of 1 percent, there is a requirement for a greater workforce than we have sustainable.

One of our challenges today is that we reward our health care systems in a very lean manner, in a just-in-time manner. And just-in-time operations means that you have only as many beds as you need. You have a lean staff to be able to minimize the investment of and time and energy and dollars in your system. And also, you carry a minimum of inventory or products that would help the patients that would be coming in.

In a just-in-time consumption and production world, the worst thing that can happen to it is an epidemic or a pandemic. Because very little demand elasticity exists in the system.

Whether it is demand for pharmaceutical or medical supplies or staffing, I think the challenge is really focusing on demand-planing tools and demand-planning models that give us a recognition of what the boundaries of the problem may be and a strategy to be able to absolve ourselves of some of the issues in a practical manner.

Dr. KADLEC. Dr. O'Toole.

Dr. O'TOOLE. There is lots of evidence that the public health workforce and our capacity to respond to the professional needs in mass casualty disasters is a mess. It just is very, very scanty.

Half of all of the Federal workforce now engaged in bio-defense is eligible to retire in the next 4 years. And that's a small number to begin with.

People do not go into public health in part because the schools of public health do not teach anything about public health practice. And I think docs, for example, are mostly doing disaster response as assistance professors and as an add-on task that is assigned to them by their department chairs.

There is, I agree, no coherent process for hospital preparedness at all. And that is something the bill has to address in one form or another.

I think the solutions to the workforce problems are very simple and very cost effective. But they are going to be costly. We ought to have training programs keyed in to requiring medical schools and schools of public health to offer practical training to people who wish to go into these areas. And there ought to be Government service pay back either at the local, State or Federal level in exchange for those kind of scholarships.

We ought to also make it a lot easier for more senior people who have experience under their belt in medicine or public health to come in and out of the Government, which would require personnel changes in the Federal workforce and would help everybody. It would acquaint the medical and public health systems with what Government is really up against and who they can look for an allies. And it would also help, I think, the workforce in the Federal work place.

Dr. KADLEC. Elin.

Dr. GURSKY. Thank you.

Thank you for the opportunity to speak here today, and I look forward later on to be talking about the system as a whole.

I agree with Dr. O'Toole. We have a number of very serious public health system issues that we need to discuss, not the least of which is the public health workforce.

I concur that we most definitely need to incentivize people going into public health as Dr. Devlin has discussed, but I think it is long overdue that we develop a credentialed professional public health workforce, and would suggest that merely incentivizing people to take public health training without requiring to get some kind of license or certificate, would be a very shortsighted approach.

Our tatoo artists and our hairdressers in this country need to have a license and have to have proof that they are competent to do what we expect them to do. And we should have no less a standard for people who are very much responsible for life and death decisions, for risk communications, for releases of vaccines to medical facilities for postexposure.

So the treats that we have been facing in the past two decades, the threats that are coming at us from African rain forests and strange looking cats and whatever else Mother Nature and terrorists have to throw at us, must be met by a workforce that is educated, not just trained.

As I look at some of the anecdotal reports of the past few years, the bioterrorism funding, I see a lot of training courses. And it is my opinion that what we are doing is not just training people in new pathogens and bioterrorist threats. We are actually training people in some of the fundamental skills they should have in public health practice.

And in fact, because we are not assuring governors that funding is sustainable, we are not really building into our public health workforce. We are hiring people who are short-timers and people who are contractors. So the turn over is great, and then we wind up having to train over and over again.

So I think in terms of the potential catastrophic ramifications to the health of the country and costs, the most cost effective measure we can make right now is to look to the year 2010 and say that people graduating from public health schools who wish to practice must have a certain standard of education and a license.

Dr. KADLEC. Chuck.

Mr. LUDLAM. Just two quick points. The 9/11 Commission said that our failure on 9/11 was a failure of imagination. And I think we have a total failure of imagination regarding public health, consequences of an infectious disease outbreak or a bioterror attack.

Talk to the officials in Toronto and China about what happened with SARS. There is not anything we would not throw at this problem in terms of resources, training, incentives or everything else in this country. Beijing, Shanghai, Hong Kong, they were closed down for several months.

Now what if we close down DC, New York, and Boston for a few months?

There was a death sentence on the books in China for the willful spreaders of SARS. The first thing they lost were some of their hospitals because people go to their hospitals.

You can lose your hospitals for all of the other things that hospitals are doing. And the worst situation is if you do not have medicines to treat people with. And you say, I do not know what it is. It is whatever you are going to quarantine. And that is the absolute collapse of the health care system.

In the center of attack, they said, get your ass on Cipro and stay on it until you get off and you will be fine. And there was no panic.

And we tell them we do not know what it is, that we do not know how to treat it, that we have no vaccine, we have no therapy. We do not even have a diagnostic. We cannot separate the worried well from the people who were exposed.

We have no diagnostic for small pox right now. We've got to wait and see whether they break out.

OK, the last issue is command and control. There is no consensus on who is in charge federally. It is absolutely a muddle created largely by the Congress between DHS and HHS. And this committee and the Homeland Security Committee have to work this out in this bill.

We have got to know who is in charge, and they have to have resources.

NDMS, the administration proposed to cut funding for NDMS. MMRS is a complete mess also. We have hardly any boots on the ground to send. We have no way to coordinate them with the State and local officials, who are the principle work forces in this area.

We are probably going to have to deploy the National Guard. They are totally unprepared for this.

We are not taking this seriously in the slightest.

Dr. KADLEC. Dr. Bacal, I think you have a question.

Dr. BACAL. I am Kira Bacal from Senator Hatch's office.

I would like to build on what Mr. Ludlam and Doctors Devlin and Gursky's comments and ask we have often, particularly and appropriately this morning, tended to focus our discussion on biological threats be they weaponized, small pox or new diseases like SARS. But I think it is also important to remember that our public health system must also respond conventional attacks such as mail bombs in Israel, shopping malls or car bombs in Oklahoma City.

And particularly, after last week's attack in London, subway as well as Secretary Chertoff's comments about reorganization as Chuck was saying of the Department of Homeland Security.

I would be very interested in hearing some remarks and some comments about what our guests feel should be the template for our Federal, State and local authorities to work together on both conventional as well as biological crisis.

Dr. O'TOOLE. If in the event of a explosive device or a car bomb, even a very big explosion—let's leave out nukes and rads for the moment—the real response is going to be the medical system, not the public health system, particularly immediately. And that has been a neglected piece of homeland security preparedness in general that really needs a lot of attention.

The problem is that that system is disconnected from government and highly fragmented. And as was said earlier, very, very pressured financially and just operationally. So it is going to be very difficult to fix.

I think the way forward for dealing with mass casualties situations is we are going to have to create regional consortia of hospitals for starters. No one knows who the organizing authority to put that consortia in place is going to be. For the most part, mayors and governors have not been interested in messing in those worlds. But some organizing authority to ensure hospitals could cooperate and collaborate is going to be absolutely essential.

And then we are going to need to spend money on thinking through how to get them prepared in a cost-effective way.

One critical aspect of all kinds of responses is going to be the ability to have situational awareness both in the medical care system and in the public health system. And the key component of that is being able to move information from the hospitals to public health. So public health has a broader view of what is going on. Is this one anthrax case? Is this 100? Is this one hospital? Is this every hospital in my region? Is everybody overflowing with burn victims, etc, etc?

We need to start drilling down with great specificity on what aspects of the huge spectrum of public health and medical response we really think are the most essential and start building those systems. And they are systems.

I would offer that the medical and the public health systems—and I have been part of both—are so broken and so stressed that there is no hope of tweaking these systems and getting them into good-enough-to-go fashion. We are going to have to start building whole new systems and committing ourselves to that.

Dr. KADLEC. Pardon me, Dr. Devlin and then Dr. Gursky.

Dr. DEVLIN. One of things that happened well after 9/11 is that the funding for public health preparedness and the funding from CDC and then the funding from HRSA for the trauma system, hospital preparedness, did come to public health.

And we have done, if I may say so, a very fine job in North Carolina of integrating those two systems so that we do have the specificity that you are talking about. We can tell you what beds are available and move people around in terms of mass care.

And then homeland security, of course, has funding for equipment that supports hospital work. So it is integrated in some States. And we are working on it in all States. If I might, I would like to hand out North Carolina's story. I will not go through it, but I will say that what we see the biggest issues are is that we do have to consider that this all threat, all hazards, and that we are dealing with radiological, nuclear, biological, chemicals. And we are dealing with hurricanes. And we are dealing with explosives. And that early detection, surveillance are absolutely critical. And that it is going to take a strong workforce at a local level, a regional, State, a multiState approach as well as the Federal response, because these are not respecters of any kind of boundaries.

So we understand this. And so the other notion is that, we talk about certainly we want to do everything we can to save lives. And we are going to have some devastating epidemics. I mean, we are at 36 years since our last flu pandemic. It is time.

So there is lots of national things that can wrong too. But the reality of it is, we are not funding the public health workforce and effort in a sustainable way.

Last year we had redirected $39 million for the cities readiness initiative. Not that we don't need to protect major populations, but we are all at risk, the school in every community, the nuclear power plants and we could go on and on.

This year the administration recommended a $130 million reduction or transfer to the SNS but taking it from the State and local public health infrastructure. The Senate has concurred with that, it is my understanding. The House did not.

But we are not putting everything that we can out there to protect, to detect, to put in place interventions that are going to literally save lives in addition to the industry issues that are here today.

So I thank you for your question and for yours.

Dr. KADLEC. Dr. Gursky.

Dr. GURSKY. If I may, I would like to move a little bit from workforce to systems given the question that is on the table. Because I think systems is the issue we have to address.

We do not have a public health system in this country. We have 50 State and 3,000 local health departments who work very, very hard on behalf of their citizens and has been extremely taxed over the past 25 years by new pathogens, clearly with no end in sight.

We are long overdue for rethinking the need for the building of something that works systematically in this country for public health.

And the question I frequently raise is what does a 21st century public health system look like?

We have already discussed one of the issues, which is workforce. So you know my thoughts on that.

Having an educated workforce that can respond quickly, agilely to new threats, to be able to make the decisions needed to deploy stockpiles and to provide risk communication messages.

One of the things fundamental for this would be an information sharing system, a health intelligence system. We have seen evidence of building parts of information systems now for a lot of years, on top of which we are building early warning and detection systems.

This should not be rocket science. In fact, we know how to build information systems pretty darn well. What we have done is give a lot of funds to our States and special cities to build systems without giving them requirements first, without providing the standards and the architecture.

If, indeed, we want to be serious about dealing with 21st century global threats, we have to build a system where we can rapidly share information, the situational awareness that Dr. O'Toole discussed, the connectivity between the public health medical and hospital sectors, as well as intelligence and law enforcement. That is absolutely critical.

The third thing we have to decide is, indeed, who is responsible for what. And as we saw on anthrax, it was very unclear if it was local, State or Federal CDC responding to events. We have that same problem in terms of anticipated relief at this point. I speak with people, public health professionals who feel the National Guard is going to be deployed to help them. CDC is going to be deployed and come to the rescue of multiple cities at one point in time. The Department of Defense is going to be deployed.

And maybe these are all possibilities, but our assumption of who is going to be involved and who is going to be in charge in a large mass health disaster has got to be decided so that we can plan efficiently and so that we can use our preparedness dollars efficiently.

Putting preparedness dollars in a community that has a public health staff of 13 people and allowing them to hire a 14th person is not going to make them more prepared.

We have to look at 21st century solutions. We have to look at what Dr. O'Toole was discussing in terms of regional approaches and bringing all of the potential responders and important stakeholders to the table and deciding what the response paradigm is going to be.

Dr. KADLEC. John.

Dr. CLERICI. I think as Chuck pointed out, this issue of preparedness is one that is actually driving the decisions on countermeasures being purchased. And we've seen this first hand. The issues need to be dealt with together.

Mr. Barrett in Chuck's exchange reminds me of one that President Bush and Senator Kerry actually agreed on in the debates and that was that the number one threat facing America is nuclear detonation.

A few weeks ago Governor Keane and Lee Hamilton both said that it is not a matter of if, it is a matter of when there will be a nuclear detonation on U.S. soil.

And yet, as we sit here today, there is nothing in the stockpile to treat acute radiation sickness. And the reason why is the Government is waiting for a cure that can be delivered within 12 hours via push-pack. That is not going to work. That is not going to treat the victims of a nuclear blast that are going to be downstream of the plume that are going to be exposed to radiation. And we do have something at hand that can be purchased. It just cannot be distributed in a way because the system is broken in order to get it to the potential victims in time.

Interesting the way the Government relies on the fact that they've stockpiled drugs such as Neopugen. Neopugen has to be administered in a hospital by a doctor. There are not enough hospital beds in America to treat the potential victims of a single nuclear blast in any major city.

So until these two halves of the equation are talking to each other, we are not going to be protected. And the countermeasures that are going into the stockpile are not going to help us.

Mr. LUDLAM. Just to follow up on the same point, we obviously did pass BioShield I. So we are supposed to be on the road to procuring a few things.

I have to say that the administration of BioShield I, which I basically wrote, is grotesque. One, it is not being used, you know, about one tenth as much as it should be. They do not have the long list of things they are trying to procure, including the radiation countermeasures. So the industry has no idea of what is the 5-year plan under BioShield. How many products are you going to procure and give us some warning.

They are asking the companies to have an IND before they bid. The company would have to preposition themselves with an IND. That takes lead time of a year or more to get ready to even bid on a BioShield contract.

So there is no sort of scheme here. Are they going to have 20 products or five products or 80 products? Or what is it? Nobody knows.

The few procurement that they have put out, I mean the anthrax RFP procurement, was utterly and totally bizarre. I mean, this is the disease that killed people in 2001. We still do not have any effective countermeasure for late-stage inhalation anthrax. Cipro does not help you. At that point the toxins start going, you are dead.

OK, they're trying to get a therapeutic. Fine. But put out a procurement that says, we'd like 10,000 to 100,000 courses of this product. And the industry said, 10,000?

Well, I mean if somebody dropped a pound of anthrax in a tall skyscraper even, it would infect 10,000 people, let alone if they did it upstream, upwind in Bethesda.

Then they said, we will take your product and we will test it in some animals, and we will not tell you what animals we will test it, and then we will get back to you, because there are several animal models in the field. And the industry was not sure which animal model would be used or how they would be testing it.

Now they are still dickering on it and still adding clauses to the contracts. I am not sure if they will ever meet with any company that will play.

Now this is leaving the public health system without countermeasures which we probably could get, you know, in a reasonable period of time. Not all of them but a few of them.

Dr. KADLEC. Tara.

Dr. O'TOOLE. I think we need a lot more focus on systems and less on stuff. And let's talk about what we might be able to do.

I would urge you to consider something radical. For example, perhaps we should require all governors to create a conduct of operations plan with doctrine and assigned responsibilities for dealing with a mass casualty event. And if they do not and if it does not get approved by HHS, they do not get a certain percentage of Medicare funding.

I mean, obligate them in some way that really conveys the seriousness of the situation.

If you do that, let me tell you that will get the hospitals to the table real quickly if they think they are suddenly going to be told what to do by their commissioners of health.

I think also that we need to talk with much more specificity and clarity about the public health system, versus the health care deliver system. They are not the same, and they are not even connected. And right now, when we say public health preparedness, the health care delivery system assumes that does not include them.

I would suggest shifting some of the operational responsibilities for epidemic response, in particular, such as giving out vaccines and antibiotics to the medical care delivery system. I think they are going to be better at it, and it would depressurize public health.

North Carolina may have this under control, but a lot of the big States just are not going to be able to do it with their public health system.

I also think there are innovative things we can do for distribution. For example, a lot of the flu vaccine in this country gets given out by Giant and Super Fresh and Costco. And we have done research on this recently. All of those entities are willing to participate. They don't know how to connect with the public health system, which is frankly too busy and overburdened to reach out to the private sector. But innovative ways of doing massive distribution I think are available, but somebody has to be the catalyst.

We are being too global and too general in our approach to public health preparedness. We need to really drill down on what systems and what capabilities we need in the next year and what we want to build for 5 and 10 years and start off all of it at once. But we have got to get very concrete and very specific. And you have got to assign stuff in ways that are meaningful.

Dr. KADLEC. I was just going to ask, and again not to truncate the discussion here, but ask a follow on question to the system question, and that really involves, if you will, the food safety piece and the ag piece. Because there is an entirely different set of systems, but yet overlapping and the public health community seems to be the nexus for that.

I was just wondering what is the state of affairs, if you will, in the good safety and ag piece of this that give us confidence that the agricultural world, the traditional veterinarians and ag health community is connected with the public health and the medical care community.

Dave.

Dr. FRANZ. Just to follow on a point that Dr. Gursky made in that regard, and Dr. O'Toole as well, talking about communication among the various sectors, medicine and public health, for example, I think, it is very important. In one portion of the ag threat, we have that kind of communication with the veterinarian and agricultural sector as well.

I split the threat really into two pieces. One is preharvest, and one is postharvest. The postharvest one is the one that we are talking about here, and this is adulteration of the food chain that eventually affects humans.

It is a public health problem much like bioterrorism is a public health and medicine problem.

And that is where the communication has to occur because many of those are zoonotics or they are food adulteration issues.

Then there is the other sector of the agricultural threat that I think we have to parse out because it is strictly an economic problem. Because of the way we value human, animal and plant life, and we should, when animals or plants are injured in some way, that is an economic problem. And that one is fairly easy to sort out actually because there are a lot of small kinds of effects that can occur which involve millions or tens of millions of dollars much like what occurs when we have a high-path avian influenza outbreak in our fighting roosters in California and it spreads to commercial flocks or when we have an outbreak of soybean rust or something like that.

Those are things that our systems are resilient enough to deal with.

There is one big outlier in agriculture with regard to economics and that is foot and mouth disease. And we have to separate that one out and look at it very carefully, specifically as we have done with small pox and anthrax on the human side.

So I think as we look at all of this, we need to lay that out and see that there is a public health piece to the ag threat and an economic piece to the ag threat, and clearly the public health piece because of the zoonotic issues has to be carefully integrated with what we have just been talking about.

Thank you.

Dr. KADLEC. I was going to go to Chuck. I wanted to come back here, but we will just come across.

Mr. LUDLAM. I think this zoonotic disease issue is obviously critical. SARS is zoonotic, avian flu is zoonotic, ebola is zoonotic, West Nile, malaria, TB, they're all zoonotic diseases.

They go back and forth to animals.

Some of them will be obviously chronic diseases. I mean, maybe the avian flu becomes a chronic disease or SARS and we can never get it out of our populations because they are wild flocks, for example or they are in pigs or whatever.

So and in market failures for developing products to treat these diseases in animals, similar market failure that we have for treating diseases in humans.

Surveillance issues, very similar. Obviously, surveillance, we ought to do surveillance in zoos. I mean, that it not quite the same as surveillance of human beings, but we need surveillance in zoos. We need surveillance in natural populations or in agriculture populations.

So I think that this is actually a critical part of the bill.

Now to be clear, the ag title is not in the jurisdiction of this committee. The tax title is not. The IP title is not in the jurisdiction of this committee. Command and control is only partially in the jurisdiction of this committee. Decontamination, this is EPA, is not in the jurisdiction of this committee. You get into export licenses, that is in the Commerce Committee. We have accounting issues dealing with the pediatric vaccinations; that is in the Banking Committee.

S. 3 doesn't have ag at all. Obviously, BioShield II has got all of these issues.

Now the big question on any of the issues we are talking about is how do we get to a comprehensive bill?

This committee, I think, could not be exercising better leadership, more decisive leadership, more visionary leadership, but you have about one-tenth of the jurisdiction that you need.

And the only person who can move us to a comprehensive bill is Senator Frist, the only individual anywhere, who wrote the book called, "When Every Second Counts," his book about bioterrorism. And every second is now counting on whether he will lead and bring all of the committees together. Because what I fear is he will hang the HELP Committee out to dry and not bring in the other committees. And you'll report out a bill. And Senator Burr and Senator Enzi will be managing a bill that will be an embarrassment, a trivial bill, another incremental, ineffective bill.

And the only way to get the help of the other committees is Senator Frist.

So the entire success of this effort depends on him and ultimately also the administration to work with him.

So that is the challenge that we have here is that we don't have leadership at the top yet. We have got it here in decisive ways, and we do not have it where it counts with Senator Frist and the administration yet.

Dr. KADLEC. Dr. Devlin.

Dr. DEVLIN. I was just going to say two points. One is that I am going to respond to the food security issue, but just that I just want to make the point that we have a public health system that is Federal, State and local. The investment is 3 year's old in asking public health to assume their responsibilities on the frontline with the police, fire, crime control and public safety.

We have come an incredibly long way in this effort. And we really need to sustain that. And I am sorry things are not good in Pennsylvania, but I am aware of many, many States that are just doing outstanding networking with their medical colleagues, with their agricultural colleagues.

We have a one-medicine approach with our veterinarians in North Carolina. It is all the same.

Now I had the opportunity to participate at the Federal level on the Government coordinating council and the industry coordination council as we try to bring—I am switching to food safety now—bring Government and the food industry together to talk in open communications. And we have a farm to fork approach. And we have got to talk about the threats. You know, they are an industry that we regulate. There is some discomfort there about telling us where their vulnerabilities are, or accepting our insights into where we think their vulnerabilities are. We have got to communicate. We have got to assess the threats. We have got to harden the different nodes that are vulnerable along the farm to fork continuum. We have to got to share out data and have a common multithreat data base that is automated, that we share so that we have good surveillance, farm to fork, of what is going on in the food industry so that we can again detect something early and make an appropriate intervention.

So we have to have a plan. We have to exercise it. And we have had wonderful cooperation. We actually fund veterinarians through our terrorism initiative in agriculture. We use their trucks. They use our radios. We use our command centers together.

We have a very strong partnership with agriculture. Of course, we are a $62 billion industry in North Carolina. Twenty-two percent of our income is agriculture. So it is a very critical infrastructure, not just in North Carolina but of course in the whole country.

So it is something that we have got to get more serious about. And we have not funded agriculture at all. We have funded food security at all. What we have been dealing with for the last century is food safety—are hot foods hot and are cold foods cold.

We do not know about the driver that brought the food in the back door and whether the back door was locked before they got there.

So there is a lot we need to do around food safety and security. We are just beginning.

Dr. POURNOOR. 3M of course is a global company. In January I launched an avian influenza campaign in Asia. There are a number of countries that are approaching us asking us to help them with getting themselves prepared both from a public health as well as a hospital perspective. And I think that speaks to some of the infrastructure questions that came up.

One of the very boring topics that nobody seems to focus on is supply chain issues. It is not as flashy as developing pharmaceuticals, and it is not as exotic as some of the other things we emphasize, but supply chain, in essence, feeds our hospitals, our public health systems to be able to sustain themselves as they respond to all sorts of emergencies.

One of the facts that we often gloss over is the situation where today, as we manage all of the goods and services that we need to deliver the care that is required, these goods and services partly are produced in the United States and partly abroad. If there is an epidemic or a pandemic, how is global trade influenced and how can the flow of goods and services from various countries to and from the United States affect it? I see that as something that, for instance, is not touched upon.

I think we have the wherewithal and the knowledge amongst folks in this room and beyond in being able to put good demand plans together. Logistics and demand planning are sort of two critical links in the chain. Demand planning requires that our intuition about what happened in an event be brought down into quantitative, tactical and operational detail.

We have developed some demand planning tools that we use to work with first responders and public health officials in trying to help them, but to go from very generic statements that believe at generic levels because we want each State to have the autonomy to implement them the way they wish, is a long way from making those generic statements about preparedness to actually having tactical and operational plans. And not everyone is taking those statements of preparedness, minimal levels of readiness, and implementing it in the same manner.

Implementation—the devil is in the details—and demand planning and logistics I believe is one of the key pieces.

Dr. KADLEC. If I could just maybe leverage what you just said as a last question, and to clean this up as we get to the close of this very informative roundtable is really kind of open in terms of what is the view concerning our status or State of preparedness for avian influenza? And I think you touched on that, and maybe we will be here for another two hours.

Dr. O'TOOLE. [Off microphone.] [Laughter.]

Dr. KADLEC. Yes, really. It is all relative, Tara, as you know.

Mr. Barrett.

Mr. BARRETT. I think the thing that John spoke to about the, as he described it, boring issues of execution and supply chain, are really in a way the low-lying fruit. And while the conversation that we had in the earlier part of the day is very important, and obviously, we need to work on both these dimensions, I think there is so much here in the flow of information, the flow of product, the flow of services that can be dealt with.

Again, there was a comment earlier about we need to take extraordinary action. All true. We also need to take extraordinary action on this very mechanical part of the process, and that is the part that at times does, again, from my perspective, seem a bit absent, and it is really this detail that I think could bear some fruit for us.

Dr. KADLEC. And again, could I just solicit some comments about avian influenza as we kind of wrap up? Elin?

Dr. GURSKY. I think we need to leave here with both the strategic issues and the tactical, and they are both equally important, and I cannot afford to not say once again what we want the 21st century public health system to do.

And we have heard from Dr. O'Toole and Dr. Devlin, and these both have validity. I think it is a matter of what we wish to invest, what our expectations are, and from that decision we clearly have to look at the tactical issues because right now we have 3,050 different approaches to delivering risk communications, stockpiles, exercises, and it is really a recipe for disaster.

Dr. CLERICI. I will come back full circle, Dr. Kadlec. Until we address the liability issue, we are not going to be prepared for a pandemic. The pandemic vaccine manufacturers will not produce product absolutely 100 percent without liability protections. They are not bluffing. You are asking them to displace their entire manufacturing process. Childhood vaccines, adult vaccines will not be produced in the event when we turn to pandemic production, and they will not play until a pandemic is addressed. And if I can take one thing away from that, the issue of pandemic liability must be addressed now.

We have a luxury, frankly, with bioterrorism because we can stop, we can arrest, we can use intelligence to prevent the terrorists from hitting us. There is nothing we can do about pandemic, and unless and until we address it, we are very much at risk. It is absolutely the scariest thing I work on every day, and the thing that keeps me up at night.

Dr. KADLEC. Tara.

Dr. O'TOOLE. I think the situation of avian flu in Asia is terrifying. I am in a State of disbelief at how little America is doing to understand the threat or to respond to it. We ought to be leading the world. We are essentially doing nothing. And I suspect that rather than bioterrorism, our supine posture and failure to act, let alone lead the world in meeting this threat, is going to be probably the most important statement on where America is and where it is heading of our generation, to sum up.

Mr. LUDLAM. Avian flu is not the last new disease we are going to see. Certainly, I mean it is one of 20 we have seen in the last 20 years, 20 new diseases. The capacity of bioterrorists to concoct new things, hybrid GM, or sophisticated chemical weapons like bioregulators and biomodulators is completely unbelievable. What the Soviets had in 1992 was unbelievable.

Now, if we do not have the ability to more quickly develop products in response to Mother Nature or things we first see from terrorists, if we cannot go from bug to drug, or bug to vaccine, or bug to diagnostic, in a fraction of the time that we now have, we are going to be surprised because we do not know what avian flu is going to look like in its final—of the human. Maybe it is similar to what we have seen, or maybe it is sufficiently different that whatever we might have developed will not work. Maybe the diagnostic will not work. Maybe the vaccine will not work. Maybe the therapeutic will not work.

So the single most important revision in every bill I have drafted on the subject is the research tool provision, saying that whatever we do in the way of creating products for known threats—and obviously we ought to knock those off in a discrete period of time with dramatic incentives to create antidotes to all of the things we know about—we need a research tool industry, including animal models and all of the rest, so that we can tear the pathogen apart and find its vulnerabilities, find out how to attack it, figure out how to develop a product, and do that in a fraction of the time at a fraction of the cost, with much more reliability than we can now.

We have to know in advance whether it will work and who it will hurt. We do not know that now. We are getting products like Vioxx was pulled after years on the market, and this is the failure of research tools. We are making terrible mistakes already. Products are dying in Phase III clinical trials or after they are approved. We need to know much more about what we are creating, how it is going to work, if it is going to work, and what the side effects are going to be in advance. And this is research power that we need in the infectious disease context like we need it nowhere else.

And ultimately the only preparedness will be more powerful research tools, assuming we have stockpiled all the products for the known diseases.

Dr. O'TOOLE. Could I say one thing on that because I do not want to leave—I agree with everything Chuck says except that I think it is very dangerous to lump avian flu in with all the other emerging diseases or bioterrorism attack. It really is special because of the way it spreads, the fact that you are contagious before you are symptomatic, and because of the immunological naivete that is definitely going to be there whatever the final bug is that ends up spreading in humans, if in fact that happens.

There ought to be something going on in Congress right now to look at what we could do about avian flu. I know there have been lots of hearings. My impression is that the administration says everything is fine, we are under control. We do not need any more money. That is ridiculous. We ought to actually have a list of things that we could be doing right now about avian flu. Someone in the Government has to generate that, and we ought to be doing it right now. There are things to be done.

Dr. KADLEC. Dr. Devlin.

Dr. DEVLIN. Just in conversations this week with representatives at CDC, they were indicating, in a more specific response to your question, that we are at least 3 years away from enough vaccine for the whole population and probably 18 months or so away—and I know you all have to make it, but that is what they are saying—and that we would be about 18 months away from having enough antivirals in place, which is also something that we are very short on.

Having said that, you know, the first thing that is going to happen is that we are going to have to put in the time-honored restraints on movement and place of people, and that we are going to have to be into quarantine and isolation, not holding the ball game, and looking at whether schools are open or not. So there are going to be some very dramatic things put in place very quickly that contain the spread of the infection while we wait on the antivirals and the immunizations.

But one thing I know that we were trying to be creative in how we break down the stockpile, exercise the plans, get the medicines out to the population, and the administration is considering some other innovative strategies like using the workforce in the postal system, and some pre-event deployment of medical kits. We have some great questions about pre-deployment to the populations of medicines that they may use when something else happens, or that they may lose, or they may get outdated. We believe that—I mean I have some questions about if something happens in one part of the country but you have pre-deployed your assets somewhere else, that is a real concern. So I just mention that.

Dr. KADLEC. Thank you. We will leave that as, if you will, the final word.

First of all, on behalf of my colleagues, I would like to thank you very much, to all the panelists for kind of hanging in here maybe a little longer than you anticipated, but it was of great value and will be great value to our members as we will be highlighting many of the points that you have raised here today. I can only thank you all for your travels, particularly those who have come out of State, and certainly your time. But again, unless there are any further comments or questions, I just say thank you and wish you safe travels home. Have a good day.

[Additional information follows:]

# Additional Information

## Prepared Statement of Elin A. Gursky, Sc.D.

### "What is Necessary To Build and Maintain a Robust National Public Health Infrastructure To Meet Future Biodefense Requirements?"

#### SUMMARY REMARKS

Health security is threatened by the intentional release of biological weapons and the occurrence of natural disease outbreaks; these events are likely inevitable and potentially catastrophic. Although the responsibility for communicable disease containment has historically resided within public health, today's homeland security challenges and microbial burdens exceed the capacities and capabilities currently residing within this sector.

Following the 2001 anthrax attacks, the authorization of almost $1 billion through the Frist-Kennedy bill (S. 1765) and subsequent awards and supplemental funding have provided critical support to a long-neglected public health infrastructure. However, we must acknowledge woefully limited evidence of strategy and systems that can work consistently, uniformly, and durably on a national scale in the event of a deliberate or naturally occurring infectious disease epidemic. Such an event will stand little likelihood of being confined within a single State, and will represent a crisis of national security significance.

Bioterrorism preparedness has exposed the frailty of the patchwork quilt that is comprised of the country's 3,000 local and 50 State health departments. Our public health agencies are a local enterprise, with a tradition of employing unique activities to *promote* the health of their communities and serve the needs of their local elected officials. However, the capabilities required—and in large part absent to *protect* populations include decision-support to manage uncertainty during a large-scale infectious disease outbreak; training in the use of sophisticated technologies that sustain disease surveillance, detection, and information sharing; leadership to integrate efforts with the medical, legal, and intelligence communities; and skills to coordinate mass care such as prophylaxis and vaccination.

Twenty-first century threats require a re-envisioned public health *system* that is agile, well trained, accountable, and uniformly effective across the Nation. Before investing another $3 billion, we must develop a strategy that will ensure the health security of our 280 million citizens.

---

Few experts dispute the inevitable and potentially catastrophic threat of a large-scale biological attack from Mother Nature or terrorists. The decades concluding the last century offered ominous insights into the evolutionary transmutations of emerging and reemerging pathogens such as hantavirus, West Nile virus, and HIV-AIDS fueled by ecological change, the global migration of humans and agriculture, and drug resistance. The 21st-century awoke to SARS and monkeypox; outbreaks of avian influenza, Ebola hemorrhagic fever, Marburg hemorrhagic fever, and polio can be found in areas less than a day's plane ride away. The threats to health security from terrorists by the dispersal of biological weapons such as smallpox and anthrax or by dispersal of naturally occurring diseases such as Ebola or plague are augmented through their arsenal of time, resources, and increasingly sophisticated biotechnology. Whether the pathogen is exotic and bioengineered or common, bioterrorism expands the element of intervention upon standard public health management. Bioterrorism can be a single attack with a single pathogen, or multiple attacks with multiple pathogens on multiple targets. Interventions in the attack(s), in the behavior of potential victims, and by the requirements for swiftly administered medical prophylaxis are among the challenges that must be rapidly and accurately assessed and to which a coherent response must be generated.

The majority of experts agree that it takes a special army to combat these threats—an army that understands incubation periods, the transmission of infectious agents across susceptible populations, and strategies such as isolation, mass vaccination, and prophylaxis to control an epidemic. The army is the public health sector. The Frist-Kennedy bill (S. 1765) authorized almost $1 billion following the 2001 anthrax attacks to build public health infrastructure and strengthen our response to bioterrorism, a prescient acknowledgement of the critical role of public health in this war against bugs. Subsequent Federal awards and supplemental funding followed. Four years and $3 billion later we can see evidence of strong preparedness initiatives in a number of States.

However, before we spend another $3 billion, we must acknowledge woefully limited evidence of strategy and systems that can work consistently, uniformly, and durably on a national scale in the event of a deliberate or naturally occurring infectious disease epidemic—an event that will stand little likelihood of being confined within a single State and that, not which will represent a crisis of national security significance.

The responsibilities facing today's public health departments are broad. Beyond communicable diseases, health departments confront a wide spectrum of tasks that include chronic disease screening and education (cancer, diabetes, asthma, and hypertension); community outreach to seniors; family planning, maternal health, and prenatal care; dental health; injury control; and social marketing to decrease tobacco use, teen pregnancy, and violence. Moreover, public health departments find themselves increasing the level of effort they must devote to serving as a medical safety net as the number of uninsured Americans rises to 45 million. Historically, crisis management has not been a developed capability of public health. In fact, the skills and talents required to accomplish and manage uncertainty and to lead effectively during a biological attack are quite dissimilar to those needed in outreach efforts for chronic disease.

Bioterrorism preparedness has exposed the frailty of the patchwork quilt that is comprised of the country's 3,000 local and 50 State health departments. Both research data and anecdotal reports indicate that preparedness efforts have interfered with routine day-to-day responsibilities and have engendered frustration and resentment as State budget crises force cuts or curtail traditional public health programs designed to promote community health status and provide a social good. Preparedness for events such as anthrax attacks have not been embraced as a "core mission" of public health, but are perceived as usurping fundamental responsibilities of the community. We must concede several key tenets before redoubling our preparedness efforts:

- Our amalgamation of State and *local* health departments is a local enterprise, bounded by the principles of federalism and directed by the needs of governors, county managers, and mayors.
- Fifty State and 3,000 local health departments comprise a sector—not a system. There are few shared practices across that sector that can translate into a systematic approach evidenced by regional public health response paradigms, "mutual aid," or surge capacity. Public health itself is highly fragmented and represents a wide spectrum of professional interests and backgrounds.
- Confusion regarding public health authority and responsibility abound. As was seen during the anthrax attacks and remains evident still as Federal preparedness funds flow from State to local health departments, there is no consensus regarding the roles and responsibilities of CDC, State, and local public health agencies during a large-scale biological attack. Note that the median number of staff in our local public health agencies is 13. Note also that after several phone calls to the CDC and speaking with 15 different individuals, it is apparently not known how many of its 9,000 staff is deployable and fully trained for a response role in the field.

There is no terminal degree or education that defines a "public health practitioner," and a large portion of our public health workforce relies upon on-the job-training. This may well serve their health promotion responsibilities, but is inadequate to effectively address health protection and security challenges of the 21st-century.

The public health sector remains essentially disconnected from many critical partners, especially the medical and hospital sectors, creating a dangerous gap between efforts to detect a disease outbreak and assure the rapid medical interventions necessary to avert a full-scale epidemic. Historically, the majority of emerging diseases and the anthrax events of 2001–02 were recognized by clinicians in clinical settings, typically outside the realm of public health. The Nation's medical system is in crisis, with very little spare capacity with which to care for an increased number of patients. Medical facilities are largely not-for-profit businesses that acutely experience the effects of changes in health care funding and liability.

Despite the availability of exercises and short courses implemented since 2001, the public health sector and the vast majority of clinical caregivers remains untrained, inexperienced, and naive regarding the scope of a potentially lethal and unremitting infectious disease outbreak. Note that at the June 9th Library of Congress meeting led by Senators Burr, Clinton, and Lieberman, and Representative Cox, former Deputy Homeland Security Advisor to President Bush, Richard Falkenrath, stated that no public health department could swiftly distribute and administer medical countermeasures from the Strategic National Stockpile.

Prudence compels us to assess the return on our preparedness investment thus far as we proceed on a course to protect America's most critical infrastructure—its

280 million citizens—many of whom will bear the responsibility of treating the sick, operating utilities and transportation systems, assuring civil order, and maintaining our business and industry in the event of a "catastrophic" disease event.

Twenty-first-century threats require 21st-century public health strategy. We must balance public health's traditional role of promoting the health of Americans, while ensuring a critical new role protecting the health security of America. Should we re-purpose public health and remove the financial and labor-intensive burden of persuading Americans to overcome their proclivity to obesity, lethargy, and tobacco? Should we retain health promotion responsibilities at the State level and federalize the public health protection components? Should we continue to invest in the entirety of the public health infrastructure, hoping we will accrue critical capabilities for detecting and responding to pandemic influenza or plague? Should we invest in more practitioners, or in technology-based solutions like BioWatch?

Difficult decisions are necessitated by the exigencies of the current threat environment, heightened just a week ago by the bombings in London. The 21st-century demands that we build a public health *system*, an entity that responds with consistency, uniformity, and efficiency across the Nation. A number of efforts will help us implement the necessary systematic approach to disease detection, intervention, and containment.

I would urge this committee to consider the following short-term steps:

We must focus efforts and resources to build a national health security information infrastructure that connects our public health, hospital, and medical communities (and also law and intelligence). Current efforts are languishing from a State-by-State approach that has been absent national standards and requirements. Real-time response to infectious disease occurrences are a critical component of national security. Note that four cases of cutaneous anthrax went unrecognized prior to Bob Stevens' diagnosis with inhalational anthrax.

- We must rebuild our public health workforce through principles not unlike those applied in the DoD's force transformation efforts. We must strive to achieve public health "special forces" to address the war on emerging and deliberately released pathogens. Grants and loans will help recruit new cohorts of public health professionals, but we must require that those choosing to be practitioners (not researchers or academics) attain a level of skill demonstrated by earning a license or certificate. The individuals who make critical decisions about the health of populations must be subject to professional accountability as are our physicians, attorneys, and even tattoo artists and hair stylists. Schools of public health must devise specific public health practice curricula. National credentialing exams must be formulated and administered by an impartial outside agency.
- We must clearly articulate the roles, responsibilities, and authorities of local, State, and Federal (CDC) public health agencies during a large-scale public health crisis. Specifically, the horizontal connections between agencies sharing responsibility and authority in a biological attack must be strengthened.
- We must foster closer integration of roles and operations between the public health and hospital sectors, through such strategy as joint planning and funding. Hospitals must receive funding to incentivize increased training and capacity and be assured relief from liability during crisis response. Public health has authority to direct care, and hospitals have capability to provide care for victims. This linkage should be specifically supported and exercised.
- We must assess the effectiveness of preparedness through strict measures of accountability and through performance in rigorous full-scale and tabletop exercises. To fully stress and shape the public health response systems, we must avoid instances where public health writes, participates in, and then evaluates its own performance. Demonstrated competence and capabilities, not attendance at a course or tabletop, are the goals.

The job of leading the effort to protect the public from potentially lethal infectious diseases falls to public health. But 4 years after the 2001 anthrax attacks, the burden of overcoming decades of underfunding, shrinking ranks, and expanding chronic health and medical care responsibilities has hampered the public health preparedness effort. Bioterrorism has not become a core mission, and funding State and local agencies has thus far demonstrated that the sum of the parts will not make up a "whole" national preparedness effort.

Before we invest another $3 billion, we must take the necessary steps to build a 21st-century public health system! Thank you.

## Response to Questions of the Committee by Chuck Ludlam

By way of introduction, I have spent the last 4 years of my public service career on a crusade to highlight the near total lack of preparedness of our Nation and the international community to the bioterror and infectious disease threat. I was the principal author of the 2001 Lieberman bioterror bill, the 2002 and 2003 Lieberman-Hatch bioterror bills, and the 2005 Lieberman-Hatch-Brownback bill, BioShield II, S. 975. I was also the principal author of S. 3, the Republican Leadership bill. And I was the principal organizer of an international panel of 600 experts to draft these bills.

Now that I have retired from public service, I am finally free to say what I know to be true: The response of the Administration and, with some notable exceptions, the Congress to these critical challenges has been grossly inadequate. As for the Administration, it has been reported to me that a high-ranking Administration official admitted that it proposed BioShield I solely to protect its right flank when Senator Lieberman was running for President, not as part of a serious bioterror strategy. It's obvious that BioShield I was poorly calculated and the industry response to it has been to yawn. Yet, despite the introduction of S. 3 and S. 975, there is no indication that the Administration will join in the effort to enact them. As for the Senate Democrats, in crafting the four Lieberman bills, and despite extensive efforts, I was never able to recruit a single Democrat to cosponsor these bills. The reason they all give is that "the generics hate it." Finally, in terms of Senator Frist, we've seen bold words, but few discernable actions. He is, of course, the only person who can ensure that the Congress takes up a comprehensive response to these threats. If we fall short in enacting some combination of S. 3 and S. 975, it will be principally his fault.

On the day I retired from public service, June 24, I sent a "parting shot" email to my panel of 600 experts and a copy of it is printed below. It's being made public here for the first time. I am happy for it to serve as my valedictory regarding the quality of my efforts and the Congressional response.

Unfortunately, it may take overriding political considerations to drive consideration of the deadly serious public policy issues addressed in this roundtable. On June 23 Mort Kondracke wrote a prescient article in Roll Call entitled, "Avian Flu Could Become Top 2008 Issue. Seriously." He accurately quotes me as saying, "You have a fascinating conflation of presidential politics and serious substance at work here. You have three presidential candidates interested in this issue—Sen. Frist, Sen. Hillary Rodham Clinton (D-N.Y.) and Sen. Sam Brownback (R-Kan.), a co-sponsor of the Lieberman bill. [Also, Evan Bayh] Whoever is out in front will look pretty good if the worst happens. Anyone who's behind the curve will look like a dolt. There will be 9/11-style commissions all over the place and hundreds of Richard Clarkes testifying that they warned about what was coming and higher-ups didn't listen." I stand by these words. I am proud to have issued these warnings and provided this leadership.

*Question 1.* What additional incentives or other measures will ensure the timely availability of sufficient amounts of effective biodefense medical countermeasures, and is the cost of such incentives acceptable?

Answer 1. The Lieberman-Hatch-Brownback BioShield II legislation, S. 975, was developed with the active assistance of my panel of 600 experts and it reflects a consensus of that group. It proposes in 29 titles and 360 pages a comprehensive and aggressive strategy of incentives for the development of effective bioterrorism and infectious disease medical countermeasures and addressees a host of other critical issues. The cost of the proposed incentives is trivial compared to the cost of a bioterror attack or infectious disease outbreak. If we are hit with a bioterror attack, or a pandemic, and if we have not secured the development of these medical countermeasures, we're likely to see public panic on a scale similar to that depicted in Spielberg's War of the Worlds, Camus' The Plague, and Bergman's The Seventh Seal. We'll be forced to go straight to quarantines, which will be exceedingly ugly. Those enforcing the quarantines might be given "shoot to kill" orders to enforce the quarantine.

Unfortunately, we are almost totally lacking in these medicines. In the summer of 2000 the Defense Science Board found that we had only 1 of the 57 diagnostics, drugs and vaccines most needed to respond to a Bioterror attack. At the time, the Board projected that we'd have 20 of the 57 within 5 years and 34 within 20 years. But, 4 years later, we have only 2 of the 57 countermeasures; we've added a diagnostic for anthrax. At this rate, we won't have 20 countermeasures until 2076 and 34 until 2132. This list doesn't include medicines for bioterror pathogens engineered to be antibiotic resistant, hybrid pathogens (like the Plague-Diphtheria hybrid developed by the Soviet Union), genetically modified pathogens, and a host of other exotic pathogens like autoimmune peptides or antibiotic induced toxins.

To be clear, it makes no sense to focus solely on countermeasures for bioterror pathogens. We know that Mother Nature is a terrorist who will attack even if terrorists don't. We need vastly more effective medicines to cure and prevent AIDS, malaria, TB, and a host of intestinal parasites, naturally occurring antibiotic resistance (where we face a national crisis), and a host of other debilitating diseases, like Hepatitis A, B, and C—that kill millions each year. In terms of the death toll, this is a moral and practical crisis similar to World War I and II combined, yet the public policy response has been pathetic.

Because the infectious disease threat is evolving, we need to establish biodefense, infectious disease, and vaccine industries able to develop countermeasures, perhaps hundreds of them, as the threat evolves. The Administration's $5.6 billion budget for BioShield I is not remotely realistic. The procurement cost for these medicines will run in the many tens of billions of dollars and it'll be worth every penny.

We also need to establish a research tool industry that will give us the power to more quickly develop countermeasures to new threats. Ultimately, this is the only way we can respond to novel pathogens. We need to repeal the NIH Research Tool Guidelines to establish sufficient economics to establish a research tool industry and not divorce it from NIH funded research regarding new tools.

From the industry's point of view, it's obvious that the "markets" for infectious disease products are deeply flawed. I am intimately familiar with the industry viewpoint because I served for 7½ years as the principal lobbyist for the biotechnology industry. For example, I've heard many executives say it'd be "crazy" to engage in research on AIDS because of "forced genericization." BIO and PhRMA played no role in the drafting of the Lieberman bills because their members don't want Congress to enact incentives that would press them to take up research in which they have no interest. Some in the industry have told me to "shut up" about incentives they feel would press them to "risk their company."

They say, "Look what happened to Bayer," which was subject to virtual expropriation of its antibiotic, Cipro, by HHS following the 2001 anthrax attack. In fact, the outrageous actions of HHS in that case have plagued our ability to engage this industry in this research. We must have credible Administration officials state categorically that these Mafioso tactics will never ever be seen again against a company that develops countermeasures for infectious pathogens. The companies must be rewarded, not vilified.

S. 3, and even more so BioShield II, propose bold and innovative incentives to create a viable market for these medical countermeasures. These bills seek to shift the cost and risk of development of these countermeasures to the biotech and pharmaceutical sector in exchange for substantial and appropriate rewards if—and only if—these companies successfully develop the countermeasures we need to defend ourselves against an attack or outbreak. This is no windfall for the industry. Companies are rewarded for success, not subsidized for running their meters. Conveniently, this is the business model the industry prefers; the better companies all believe that is the government funds the research, the companies will receive a cost-plus rate of return, which is totally inadequate to satisfy their investors. Creating a GoCo will definitely end any possibility that we'll be able to recruit the industry to take up this research. In fact, the industry tells me that they'd welcome a GoCo because it'd let the industry off the hook. Adopting a defense contractor model, where the government assumes all the risk as in a "Manhattan" model, has been tried and proven to be the most expensive and least productive way to proceed. S. 3 and BioShield II are premised on the notion that we can and should use the biopharma industry's entrepreneurial culture to our advantage. This is the only approach that might succeed.

The opposition of the generics to the intellectual property incentives in S. 3 and BioShield II constitutes a classic and predictable NIMBY response. Its opposition is based almost entirely on misstatements about the terms of the proposed incentives and exaggerations about their potential impact on the cost of health care. It is true that there might be some increase in the cost of healthcare if bio/pharma companies assume the risk and expense of this research and successfully develop a high priority new chemical entity that we need to protect ourselves against a bioterror attack or to cure AIDS or another deadly pathogen, but this cost should be weighed against the devastating costs if we fail to secure the develop the needed medical countermeasures. In the end, the Congress must calculate the costs and benefits of the IP incentives, such as it did when it voted to provide patent extensions when biopharma firms secured pediatric labels on pharmaceuticals.

Biopharma industry representatives have told me on innumerable occasions that the "only" compelling and realistic incentives in S. 3 or S. 975 are the IP incentives that the generics oppose. They say that if we enact all of the proposed incentives in S. 3 and S. 975 without dilution, we stand a reasonable chance that we will be able to overcome the deep industry skepticism about this research. It's imperative that we do so. In Monday's Wall Street Journal Retired U.S. Army Major General

Phil Russell, a physician who until recently was a senior adviser to HHS on biodefense issues, States, "God, if Merck or Glaxo or Aventis were involved, it would make life infinitely easier. With small companies, you have to watch them like a hawk." If you want the large pharma companies to help us, you have no choice but to enact bold incentives, including IP incentives. If the Congress buckles to the opposition of the generics and fails to include these IP incentives, it is quite likely that the legislation will fail to achieve its objectives in terms of countermeasure development and we will remain vulnerable to catastrophic morbidity and mortality, public panic, and quarantines.

If you interview the officials in Toronto or China about what they experienced with SARS, it'll transform you're approach to this legislation. You'll conclude, as I have, that developing medicines for these pathogens is an unprecedented and overriding national imperative that justifies the most aggressive and innovative incentives. You will brush aside the NIMBY opposition to these measures.

The IP and tax incentives proposed in BioShield II are not, of course, issues pending in the HELP Committee. I have suggested that the HELP Committee report out a bill with the architecture for a comprehensive bill with brackets, each of which would be left blank except to say "Judiciary Committee," "Finance Committee," "Agriculture Committee," etc. (indicating where to insert the contributions of the other committees). This is the only way for the HELP Committee to demonstrate that it supports enactment of a comprehensive bill. S. 3 includes subject matter within the jurisdiction of at least 4 Senate Committees and S. 975 at least 8 committees. Only Senator Frist can bring all the committees together to fashion an appropriately comprehensive bill.

Finally, BioShield II also addresses the entrenched ineffectiveness of the NIH technology transfer program, undoubtedly the most bureaucratic and risk averse program anywhere. It proposes to strengthen the NIH approach to technology partnerships and protect the value of its patents. If this is not done, then essentially nothing that is funded at NIH will be useful at the beside to patients. The academics who receive NIH grants, represented by AAMC, oppose these reforms because they oppose holding NIH and its grantees accountable for the impact of NIH funded research on "healthcare," but this puts AAMC deeply at odds with the patient groups for whom "healthcare" is the only bottom line. Of course, it was the patient groups, not the academics, who won the doubling of NIH funding. I suggest that the NIH reauthorization be folded into S. 3/S. 975; the two are complementary and interrelated.

*Question 2.* What is necessary to build and maintain a robust national public health infrastructure to meet future biodefense requirements?

Answer 2. BioShield II, S. 975, also proposes an effective strategy for building and maintaining a national public health infrastructure to meet future biodefense and infectious disease requirements. One key issue is command and control. To be blunt, today no one is clearly in command in the event of an attack or outbreak. This issue must be resolved by the Senate Homeland Security and HELP Committees. Again, this will only happen if Senator Frist brings the committees together to fashion a comprehensive bill.

Question 3. What is necessary to protect our food supply and agriculture from biodefense threats?

Answer 3. BioShield II, S. 975, also includes an effective strategy for protecting our food supply and agriculture from bioterror and infectious disease threats. Approximately 60 percent of the infectious disease pathogens we fear, including Avian Flu, SARS, Ebola, Marburg, Malaria, Chagas, Schistomiasis, Hantavirus, and Lyme Disease/West Nile Virus, are zoonotic—they go back and forth between man and animals. Only Senator Frist can ensure that we engage the Senate Agriculture Committee.

Overall, with regard to S. 3 and S. 975, we need to act as if the fate of civilization depended on it, which is a fair characterization of the reality of the situation.

---

## "PARTING SHOT" E-MAIL FROM CHUCK LUDLAM TO 600 BIOTERROR AND INFECTIOUS DISEASE EXPERTS (JUNE 24, 2005)

This is my last e-mail to this group. It's now 40 years since my first day as an employee on Capitol Hill. Paula and I leave to start Peace Corps training in Senegal on September 25. I very much appreciate all the kindness that you have shown to me. It's been quite overwhelming. Several hundred of you helped us to write BioShield II.

It is urgent that you maintain very high expectations of Senator Frist and the Administration. The fate of this legislation lies almost entirely in their hands. Senators Lieberman, Hatch, Brownback, Enzi, Burr, and Gregg have provided superb leadership, but there are severe limits on what they can accomplish without the leadership of Senator Frist and the Administration.

Only Senator Frist can bring together all of the Senate Committees with jurisdiction over elements of BioShield II. And nothing will happen until the Administration finally states unequivocally that we need to enact something like BioShield II.

The key problem is jurisdiction. The HELP Committee has limited jurisdiction. Senators Enzi, Burr, and Gregg can only report out a bill covering a few of the subjects in S. 3 and even fewer from S. 975. I have suggested to the committee that it report out the architecture of the entire bill, with open brackets to accommodation the contributions of the other committees. This is a way to force Senator Frist to lead. He's given a sensational speech at Harvard on these issues, but it contains nothing about his plans for the legislation. This is odd for a person in his position when he has command of what the Senate will fashion as a response.

I have sent out hundreds of emails to this group. Senator Frist's staff has received them all, and so have about 40 top ranking members of the Administration. We have here a public record of the warnings that are contained in these e-mails. If they do not heed these warnings, there is no possible excuse.

For anyone who understands the potentially catastrophic consequences of a bioterror attack or infectious disease outbreak, BioShield II is a modest and minimal proposal. Those that seek to cut back on what we're proposed in BioShield II—particularly the generic pharmaceutical industry and their Senate supporters—take a terrible risk with the public health. If they succeed in limiting the incentives in BioShield II, they will bear personal and moral responsibility if we experience an attack or outbreak for which we are unprepared. Their Nimby position and reflexive hatred for the pharmaceutical industry put the nation in peril. Given the dire nature of the threats we face, the misrepresentations they have spread about the terms of BioShield II, particularly the Wild Card patent, cannot be excused as routine lobbying hyperbole. It is possible that thousands and millions might die in an infectious disease outbreak. With Avian Flu running at a 55-70 percent lethality rate, it's possible to see a billion people dying—the lethality rate of the 1918 Flu Pandemic was 1.8 percent and 20–100 million died.

To those who say that BioShield II and this email are "over the top," I am happy to let history judge. Others can take full responsibility for ignoring the warnings I have published here. They can also take responsibility for the millions who may die if we delay development of an anti-viral that kills the AIDS virus, a malaria vaccine, and a new class of antibiotics. There is simply no price that is too high to pay for the development of these critical medicines. Unfortunately, I expect in Senegal to see many of my villagers die of infectious disease and in every case I will blame the opponents of BioShield II. My only consolation is in knowing that I have done absolutely everything possible to secure the development of these medicines, with no holds barred.

In terms of the Congress and Administration, there is zero political risk from backing an aggressive set of incentives and programs. The only risk is in not taking these threats seriously enough and cutting back on BioShield II. If politicians do not lead, they also will bear personal moral responsibility for our lack of preparedness.

Thank you all again for you support. This is your legislation to win or lose. It's yours to win now. My role is over. I wish you the best and will be forever grateful for your support.

CHUCK LUDLAM.

RESPONSE TO QUESTIONS OF THE COMMITTEE BY MR. WRIGHT

*Question 1.* What additional incentives or other measures will ensure timely availability of sufficient amounts of effective biodefense medical countermeasures and is the cost of such incentives acceptable?

Answer 1. There are a number of incentives that will improve the Nation's ability to acquire effective countermeasures in a timely way. These include:

- **Transparency**—The USG should define and guarantee the countermeasure market. This includes a clear and predictable process—timelimes, deliverables, and deadlines should be articulated clearly and as early as possible. It is difficult for companies to make long-term business decisions about which countermeasure programs to invest in if there is not a clear message from the government regarding what it intends to buy, how much, and when.
- **Valley of Death**—There is a funding gap between proof of concept and advanced development including scale up and production. Successful countermeasure

development is dependent on a partnership between industry and the USG. Funds should be made available to bridge the current funding gap. DOD has experience funding countermeasures and is a good case example for Project BioShield.

- **Contract Funding**—BioShield I allows DHHS to provide up to 10 percent of the value of a procurement contract in advance payment to support development activities associated with fulfilling a the contract, however, DHHS has interpreted the law to only allow payment upon product delivery. We believe progress or milestone payments are appropriate, particularly in the absence of "valley of death" funding and the financial commitments required for scale up and production.
- **Coordination**—A number of agencies and Departments are involved in the countermeasure requirements process. These include DHHS, DHS, DOD, OMB and many others. The process is too complicated and cumbersome and it is unclear who or which department has the ultimate decision-making authority. The process should be streamlined. It may be appropriate to consider moving all of the various components (threat analyses, research, development, and procurement) to one agency or department to improve better coordination and cooperation. The Biodefense agency could be housed at DOD, which is familiar with the development of complex weapons systems, with no commercial market, or at DHHS. If housed at DHHS the research and development (NIH) piece should be much more closely linked to OASPHEP. Greater coordination between the two would allow the identification and support of promising technologies at a much earlier stage.
- **Liability**—Currently, companies must negotiate liability protection as part of the contract process. This is not a rational way to do this and it leaves companies with potentially untenable exposure for the development and use of, in some cases, unapproved products under an emergency use authorization. The USG should provide explicit liability protection for companies and provide fair compensation for those harmed.
- **Manufacturing**—Scale up and production requires extensive up front financing. Further, U.S. capacity to manufacture biologics is limited. We believe DHHS should consider as part of any procurement contract, directly funding scale up and production as has been the case with the VaxGen anthrax vaccine and the Acambis smallpox vaccine. Additionally, ex-US manufacturers should be considered for Project BioShield procurements.
- **Funding**—$5.6 billion is insufficient to adequately support the breadth of technologies needed to protect this Nation. If the USG is truly committed to a strong biological and chemical defense, it will need to commit the funds necessary to do it right.

*Question 2.* What is necessary to build and maintain a robust national public health infrastructure to meet future biodefense requirements?

Answer 2. Efforts to improve communication and coordination between CDC, local public health agencies and target providers are necessary. In our discussions with hospital officials who would likely administer our products in the event of an emergency, we found that there was a disconnect between the CDC's plans for SNS countermeasure distribution and management and how hospitals on the front line view their role. Public statements by CDC officials indicate that CDC plans to use the existing public health infrastructure to distribute SNS IND products during an emergency. Distribution would occur within 12 hours and CDC would work with local public health officials and target hospitals to coordinate protocols, screening, tracking (adverse events, response rates), safety monitoring, protocol adjustments, and liability issues, etc.). However, discussions with a key hospital in the Washington area with likely responsibility for patient care in the event of an emergency identified the following issues:

- The hospital had little, if any, interaction with the CDC regarding emergency planning and the USG's plans under Project BioShield.
- The hospital was not aware that the USG planned to stockpile and distribute (when appropriate) IND products.
- The hospital would have a very difficult time during an emergency obtaining approval from their IRB (which they would have to do) for use of an experimental product. Staff had little confidence that hospital attorneys could work through liability issues, if not vetted in advance.
- The hospital had little confidence that the local public health infrastructure could handle/coordinate distribution of SNS products.
- The hospital did not have a good understanding of the expanded nature of the SNS. They are familiar with the distribution of push packs etc., but did not view that process as compatible with what would be needed in a bio/chem. emergency.

Activities that might improve the local public health infrastructure and ensure that needed countermeasure products are provided to those who need them quickly and efficiently in the event of an emergency include:

• Instituting mechanisms now to involve target hospitals in key cities that are likely to be impacted in an emergency to participate in strategy discussions with the USG.

• Providing funding through the procurement process to support company lead efforts to educate front-line hospitals about products being purchased for stockpile after contract award. This education process would ensure hospital staff:

• Become familiar with the products—given many would not be licensed or if licensed likely unavailable for other uses,

• Learn how to administer and store them,

• Develop protocols for their use, and

• Place on formulary or whatever mechanism is necessary to allow use during an emergency, and vet any legal or liability issues in advance.

• Allowing the distribution of SNS products through established drug distributors (McKesson, and Cardinal, etc.) rather than relying on CDC and the public health infrastructure to get drugs out in an emergency. Develop these plans and execute agreements now in advance of an emergency.

• Providing small stockpiles of SNS products to hospitals in advance of an emergency.

• Coordinating with DOD to develop plans for countermeasure distribution and administration in the event of an emergency.

## Prepared Statement of Clay Elward, Benefit Plan Design Manager, Caterpillar, Inc

Dear Senator Burr, Senator Kennedy and members of the committee, thank you for the opportunity to participate in this important Roundtable *"When Terror Strikes—Preparing an Effective and Immediate Public Health Response,"* as part of the national discussion of America's readiness in the face of a bioterrorist attack or other source of infectious disease epidemic.

Caterpillar is committed to working with you and the Administration to strengthen America's biodefense response in ways that will accelerate research, development and manufacturing of novel countermeasure agents [1], as well as diagnostic and environmental warning/detection devices. We believe that this committee can and should strengthen Bioshield I by considering the addition of certain incentives, such as needed product liability protections, guaranteed purchasing, expanded tax incentives, additional Federal research dollars, and fast-track FDA review of drug and device applications.

However, as explained below in response to the specific questions posed by the committee, the provisions in the Project Bioshield II Act of 2005 (S. 975), though an admirable effort to protect U.S. citizens, will have unintended consequences in higher prescription drug costs for consumers and will actually undermine the goal of the development of novel countermeasures by merely encouraging minor changes to already approved products.

### Introduction

For more than 80 years, Caterpillar Inc. has been building the world's infrastructure and, in partnership with its worldwide dealer network, is driving positive and sustainable change on every continent. With 2004 sales and revenues of $30.25 billion, Caterpillar is a technology leader and the world's leading manufacturer of construction and mining equipment, diesel and natural gas engines and industrial gas turbines.

And like all employers—including other members of the Coalition for a Competitive Pharmaceutical Market, a group of businesses, insurers, generic drug manufacturers and others—Caterpillar is very concerned about the costs of healthcare in the United States today. Our company provides benefits that rank among the best available anywhere, covering 140,000 lives. But at a cost of more than $600 million a year and rising, that commitment comes with challenges that must be addressed. Rising U.S. health care costs—including double-digit increases in pharmaceuticals, the fastest growing category of health care costs—have become a significant, long-term competitive issue that is impacting our ability to compete with a U.S. manufacturing base in a global market. From 2002 to 2004, our prescription drug costs increased 20 percent.

---

[1] The term "novel" as used throughout this document means new molecular entities and new and modified vaccines.

Caterpillar is addressing health care cost issues at every turn. We're working hard with providers to address cost and quality issues. We established networks to capitalize on large concentrations of employees to take advantage of the economies of scale. And we've asked health plan participants to help share more of the costs. We are asking them to be better consumers of healthcare, specifically in the area of prescription drugs, and providing them the tools to do so. As a result, our plan beneficiaries are choosing generics 90 percent of the time when they're available and that is helping control costs without impacting quality of care.

Moreover, as an industry leader, Caterpillar understands and appreciates the value of innovation. Our company has received more than 2,500 patents in the last 5 years and in 2004 alone, we spent $928 million on research and development. We also understand the competitive environment and the value of a level playing field in patent protection, which is why we have serious concerns about provisions in S. 975 that would add additional patent protections and market exclusivities to the law but would do nothing to facilitate the development of these new medicines.

RESPONSE TO QUESTIONS OF THE COMMITTEE BY CLAY ELWARD

*Question 1.* **What additional incentives or other measures will ensure the timely availability of sufficient amounts of effective biodefense medical countermeasures, and is the cost of such incentives acceptable?**

Answer 1. Caterpillar believes that the Bioshield I legislation enacted last year provides a solid foundation for meeting the challenge of preparedness against bioterrorist attacks. Indeed, we believe that this law exemplifies what can result when the Federal legislative process works best—producing bipartisan legislation that uses private-public partnerships in research, procurement and contracting to ensure that our Nation has the countermeasures we need, when we need them most, without imposing addition cost burdens on America's health care system.

As Congress seeks to enhance Bioshield I, Caterpillar recommends the following additional provisions:

- *Product Liability Protections:* These necessary provisions protect drug manufacturers as they develop and produce these potentially life-saving novel treatments.
- *Guaranteed Purchasing:* The Federal Government must commit to and follow through on purchases of countermeasures in sufficient quantity to stockpile so manufacturers have predictability for their business models.
- *Research and Development Tax Incentives and Manufacturing Grants:* The majority of pharmaceutical research in America is outsourced by brand drug companies to biotechnology and other smaller companies. The expansion of Bioshield I's funding approach would energize these entities to create a biodefense pharmaceutical sector. Also needed to ensure production are tax credits, grants and consistent government funding throughout the full development cycle of countermeasures . . .
- *Support Building Facilities In America:* Domestic production facilities will facilitate additional manufacturing capacity and assure availability.
- "Fast-Track" FDA Review: Allowing the Food and Drug Administration to expedite the introduction of countermeasures to marketplace will help ensure countermeasures are available when needed.
- *Full Funding:* Pharmaceutical companies that attract investment for research and development often experience financial hardship while awaiting Bioshield procurement contracts to materialize. The time lag creates a "valley of death" for companies developing countermeasures that could be overcome by full funding.

Americans deserve strong protections against bioterrorism, but this security must not—and need not—cripple our Nation's health care system through dramatic cost increases. As Congress considers incentives for the creation of new and better countermeasures against terrorist threats, it must strike a balance between the cost of the program and the potential benefit. Under Bioshield I, Congress established a straightforward government contract and procurement model that spreads the burden of this national defense initiative among all U.S. citizens. But S. 975 moves away from that model and places the cost burden on only one segment of the population—America's sick—forcing those that are most vulnerable to pay for this Nation's biodefense pharmaceutical preparedness. And, unlike the current proposals, an appropriate model will reward only the actual production of a novel medicine designed to address a specific security need without jeopardizing the future affordability of the healthcare system. Several incentives being debated would not facilitate the development of new medicines. As Congress seeks to enhance Bioshield I, Caterpillar opposes the following additional provisions:

Broad definition of the term "countermeasures:" By broadening the definition of "countermeasure," many drugs in today's medicine cabinets—such as hypertensive medications and cholesterol lowering drugs—would qualify as a countermeasure, and would be eligible for patent extensions. The term "countermeasure" is not limited to novel countermeasures and could apply to drugs to treat indirect or secondary effects of an attack (e.g. Post Traumatic Stress Disorder) as opposed to direct harm caused by the bioterror agent. Therefore, already marketed treatments would therefore receive "windfall" benefits. Simply put—a broad definition of "countermeasure" (1) does not properly incent companies to bring new products to market to better protect the American people and (2) unnecessarily adds cost to the U.S. health care system.

Patent term extensions: Apply to prescription medicines that are already on the market—not entirely new products, and not drugs that are solely related to bioterrorism.

Current law grants market monopoly status to a brand company that holds a patent for a new, or novel, drug product. The length of this status, or patent life, is determined by two calculations: the amount of time it takes the Food and Drug Administration to review the new product, and the amount of time that the brand company conducts its own research on the product. Brands are allotted up to 5 years of patent restoration time to account for FDA's review time. They also are allotted up to a total of 14 years of monopoly status for the development of the product, although the patent life is frequently longer because of other provisions in current law, such as pediatric exclusivity.

S. 975 would lift all of the caps on this monopoly status, extending it indefinitely. The bill removes entirely the 5-year patent restoration limit for the FDA review and sets no limit on the monopoly status awarded for the overall development period. And, instead of counting the monopoly time from when the drug application is being researched and reviewed by the FDA, it begins when the product is submitted as a rough idea in an application to the Patent and Trademark Office (PTO). Thus, the monopoly would stretch for a much longer time period than 14 years.

Of additional concern, under S. 975, a brand company could conduct a small study—related to bioterrorism or otherwise—on a drug that is currently on the market and has a capped patent life. By doing this study, the product would be eligible to receive additional monopoly status based on the time that the product was submitted to the PTO. Thus, a product whose patent is about to expire could enjoy a new extended monopoly period. Even more egregious is the fact that the product only needs to be "successfully developed"—the product never has to be approved or reviewed by the FDA, nor must it be produced or stockpiled. It merely needs to be studied for a countermeasure indication.

The provisions in this bill allow brand pharmaceutical companies to game the patent system and block affordable medicines from coming to market for an indefinite period of time. At a time when consumers are struggling to manage increasing prescription drug costs, the brand name pharmaceutical industry should not be asking Congress to extend their monopolies and maintain higher prices for consumers for years to come.

Extended Marketing Exclusivity: Marketing exclusivity delays the entry of generic drugs despite expired patent protections. Truly novel medicines already receive 5 years of marketing protection under current law. Extending the length of this exclusivity to 10 years would unfairly delay consumer access to generics.

Wild card exclusivity: The wild card is a 6-month to 2-year patent extension that could be placed on any product in a company's portfolio—even a product that is completely unrelated to bioterrorism. The countermeasure does not have to be related to a bioterror agent, but could include drugs to treat non-weaponizable diseases if the Department of Health and Human Services deems those drugs to be in the national interest.

A brand company also could apply more than one wild card to the same product, thus extending the monopoly and maintaining higher prices for consumers. Brand drugs are covered by 10 or more patents, and each of those patents could be extended. This is an unreasonable and costly incentive that provides a windfall to the brand name pharmaceutical industry, while imposing huge costs on an already overburdened healthcare system.

Taken together, those provisions would inflate drug expenditures; impose major obstacles to the entry of generic drugs into the market; worsen the healthcare crisis for uninsured and older Americans who pay for prescription drugs, and impose and inequitable burden on health care purchasers. Thus, the proposed provisions in S. 975 give a blank check to brand companies in the form of a patent extension on non-bioterror blockbuster drugs.

*Question 2.* **What is necessary to build and maintain a robust national public health infrastructure to meet future biodefense requirements?**

Answer 2. Caterpillar believes that we owe it to our employees, communities and shareholders to look toward the future and make decisions today that enable continued success tomorrow. That is why we support measures that will build upon the foundation of Bioshield I to ensure America's biodefense preparedness.

In addition to the product liability protections, guaranteed purchasing and stockpiling, research and development tax incentives and manufacturing grants, and fast-track FDA review tools proposed above, Caterpillar also recommends that the Committee focus on incentives for biotechnology companies and universities that research novel countermeasures—not for entities only interested in extending current product monopolies that are already economically viable.

*Question 3.* **What is necessary to protect our food supply and agriculture from biodefense threats?**

Answer 3. When Caterpillar was founded in 1925, we offered only one product: the track-type tractor. The machine quickly became our flagship product and a favorite among farmers because of its reduced soil compaction. Although Caterpillar does not manufacture agriculture specific products today, we do supply the industry with components including diesel engines and drive trains. In addition, farmers continue to find multiple uses for Cat equipment such as skid steer loaders on farms—large and small—across the United States.

Caterpillar supports the Unites States Department of Agriculture (USDA) in its biodefense readiness endeavors. Through their coordinated efforts, the USDA can focus on key areas of America's food supply and agriculture production. Caterpillar endorses the USDA's focus on protection through prevention, for example avoiding the introduction of agriculture health threats at our borders and reducing the opportunity for disease outbreaks and pest infestations among our farm animals and crops. In addition, the USDA's efforts to increase laboratory capacity for testing to identify hazards, such as biological agents, and monitor food-related consumer complaints will help ensure American's enjoy a safe supply of meat, poultry, and egg products.

Conclusion

Caterpillar thanks the committee for the opportunity to participate in this critical national debate. Given the high cost implications for all involved, it is Caterpillar's hope that policymakers will adopt means to assure the safety and security of U.S. citizens without jeopardizing the future affordability of our health care system as done by the above-identified provisions in S. 975.

We look forward to continuing to work with the committee and the administration on this matter.

Summary

Caterpillar Inc. is honored to participate in the Senate Health, Education, Labor, and Pension Committee's Roundtable, entitled "*When Terror Strikes—Preparing an Effective and Immediate Public Health Response.*" The discussion, to be held July 14, 2005, will help Congress put measures in place to advance America's readiness in the face of a bioterrorist attack or other source of infectious disease epidemic.

As a global business headquartered in Peoria, IL, Caterpillar supports the work of Congress and the Administration to strengthen America's biodefense response. Of primary focus for Caterpillar are the implications of bioterrorism preparedness activities on the U.S. health care system. Clay Elward, Benefit Plan Design Manager, will represent Caterpillar before the panel.

Currently the Senate HELP Committee is considering the Project Bioshield Act of 2005 (S. 975) as a means for improving protections of the American people from bioterrorist attack. Caterpillar supports much of the proposed legislation, including provisions, which will accelerate research, development and manufacturing of novel countermeasure agents, as well as diagnostic and environmental warning/detection devices.

However, while the company applauds the intent of the legislation, it has concerns with particular provisions of the bill.. If implemented in its current form, S. 975 could produce the unintended result of higher prescription drug costs for American consumers, reduced pharmaceutical access for the uninsured and added strain for the delicate U.S. health care system.

We believe that Congress can—and should—implement additional biodefense legislation to help protect the United States, but in so doing, it must strike a balance fair to all Americans.

About Caterpillar Inc.

For 80 years, Caterpillar Inc. has been building the world's infrastructure and, in partnership with its worldwide dealer network, is driving positive and sustainable change on every continent. With 2004 sales and revenues of $30.25 billion, Caterpillar is a technology leader and the world's leading manufacturer of construction and mining equipment, diesel and natural gas engines and industrial gas turbines.

PREPARED STATEMENT OF JOHN POURNOOR, 3M COMPANY

Mr. Chairman (and members of the committee): I would like to thank you for calling today's round table on "Preparing an Effective and Immediate Public Health Response," and for inviting 3M to share its experiences and perspective in this area.

3M is a diversified global technology company with international operating units in 65 countries and more than 67,000 employees worldwide; roughly one-half of our employees are located in the United States. 3M's worldwide sales in 2004 were $20.0 billion, of which 61 percent—or $12.1 billion—were international sales outside the United States. Of note, exports from our U.S. plants were a critical component of our international sales: In 2004, 3M exported almost $3.8 billion in finished and semi-finished goods that were manufactured in our facilities in the United States. This ranked 3M as the Nation's 39th-largest exporter in 2004, up from the 50th-largest exporter in 2003, giving 3M an almost 4-to-1 trade balance in favor of exports. 3M also annually invests more than $1 billion on research and development. We manufacture over 50,000 products and are world-class producer of respiratory protection products, medical supplies, and food microbiology solutions among many other categories.

Mr. Chairman, I chair CBRTA, an alliance of industrial, non-profit and academic institutions successfully leveraging our own investments in R&D with both accountability and IP protection to more rapidly prototype needed government solutions. CBRTA focuses on chemical, biological and radiological solutions. I have facilitated State homeland security exercises, and have worked with our teams on our Public Health Solutions initiatives, and recently launched, with our international teams, our Avian Flu preparedness campaign in the Asia-Pacific region.

3M works with many local, regional and national agencies and institutions on fulfilling requirements for emergency preparedness and response. Our products and service offerings help local and State governments in areas of patient surge, isolation, registration and credentialing, stockpiles, personal protective equipment, decontamination, triage and trauma, information technology, education, training and preparedness exercises. My role has placed me at the crossroads of needs and capabilities in certain areas of Homeland Security, Defense and Public Health. It is from this tactical and operational perspective that I would like to share our perspective with members and participants in this forum.

Mr. Chairman, today, the United States is investing in a variety of national preparedness programs stretching in outcomes from the development of new vaccines, to stockpiles of pharmaceuticals, personal protective equipment and medical supplies and many others in order to raise our levels of readiness in response to natural or man made bio-events.

Because we operate in a just-in-time and lean manufacturing economy that also applies to health care delivery systems, little supplies inventory exists to respond to a sudden surge of patients for threats like epidemics, pandemics or mass casualty events. Accordingly, the timely availability of effective bio-defense medical countermeasures requires that first, projections of potential patient loads be made, and then proportional demand plans be put in place to respond to such patient loads. NIH, HRSA and CDC are stimulating and fueling consideration of preparedness levels and augmentation of the system with needed caches. These demand plans must address adequate supply, purchase and distribution of needed medical countermeasures as well as stratification of priority groups receiving care.

The wrinkle in this new era of public health demand planning is that not only both the local characteristics of the health care networks and the epidemiology of the event must be considered, but supply chain and logistics factors must also be incorporated. Supply chain and logistics is often not viewed as a function of public health. Yet, it happens to be one of the core competences of U.S. industry and a requirement for effective public health surge-response. This suggests an opportunity for public-private pre-event planning and partnership assuring uninterrupted flow of needed goods and services during a bio-event. 3M is developing unique demand planning tools aligned with each of these principles in such areas as health care personal protective equipment, decontamination, medical supplies, and mass clinics.

One can zoom out from the view of flow of needed goods to local and regional health care systems to the global economy and the flow of manufactured goods across borders. In many cases our consumption demands are met with production both at home and in other parts of the world. Uninterrupted flow of these goods and services requires that we assure continuity at a global trade levels during worldwide bio-events. This makes it even more imperative that, today, during the pre-event period, we build a cushion for a future surge in demand in the system. Surge in demand during an event will cause a surge in production and a consequent surge in needed capacity and needed raw materials. These core elements of good manufacturing practices are to now be also viewed as fundamental elements of an effective public health response.

Aside from preparing for greater supply-demand elasticity during a surge in needed health care resources, measures can be taken to stimulate development of new solutions in response to new challenges and threats. Some of these opportunities have or will be touched upon by other participants at this round table discussion.

Creation of incentives to leverage commercial investment in technology towards developing new solutions is key imperative. The incentives for such a leverage span from continued R&D funding of government-industry partnerships from small businesses to large ones, clear articulation of risk management and indemnification boundaries, protection of intellectual property rights and ultimately development of sustainable and practical business models around bio-defense.

We know intimately how R&D expenditures—and the protection of intellectual property assets—can spur innovation. Last year for example, 3M received close to 600 U.S. patents—a direct result of our $1 billion-plus R&D investment. In the intellectual property area, patent reform, prior user rights, and research exemptions can play a significant role in reducing risk of R&D investment. In this area a good place to observe lessons learned is the Orphan Drug Laws. We believe creation of analogue tracks to spur commercialization of bio-defense solutions is appropriate.

3M and I are thankful for the opportunity to provide input and work toward solutions with you. I hope I was able to touch on a few key topics from a manufacturers perspective in the time allotted to me and will be happy to answer any questions you may have.

[Whereupon, at 12:20 p.m., the Roundtable was adjourned.]

○